An Introduction to Science and Technology Studies

Sergio Sismondo

© 2004 by Sergio Sismondo

350 Main Street, Malden, MA 02148-5020, USA
108 Cowley Road, Oxford OX4 1JF, UK
550 Swanston Street, Carlton, Victoria 3053, Australia

The right of Sergio Sismondo to be identified as the Author of this Work has been asserted in accordance with the UK Copyright, Designs, and Patents Act 1988.

First published 2004 by Blackwell Publishing Ltd

Library of Congress Cataloging-in-Publication Data

Sismondo, Sergio.
    An introduction to science and technology studies / Sergio Sismondo.
        p. cm.
Includes bibliographical references and index.
    ISBN 0–631–23443–8 (hardcover: alk. paper) – ISBN 0–631–23444–6 (alk. paper)
1. Science – Philosophy. 2. Science – Social aspects. 3.
Technology – Philosophy. 4. Technology – Social aspects. I. Title.

Q175.S5734 2004
501—dc21

                                                                    2002152570

A catalogue record for this title is available from the British Library.

Set in 10 on 12.5 pt Galliard
by Ace Filmsetting Ltd, Frome, Somerset
Printed and bound in the United Kingdom
by MPG Books Ltd, Bodmin, Cornwall

For further information on
Blackwell Publishing, visit our website:
http://www.blackwellpublishing.com

Contents

Preface

Science and Technology Studies (S&TS) is a dynamic interdisciplinary field, rapidly becoming established in North America and Europe. The field is a result of the intersection of work by sociologists, historians, philosophers, and anthropologists studying the processes and outcomes of science and technology. Because it is interdisciplinary, the field is extraordinarily diverse and innovative in its approaches. Because it examines science and technology, its findings and debates have repercussions for almost every understanding of the modern world.

This book surveys a group of terrains central to the field, terrains about which a beginner in S&TS should know something before moving on. For the most part, these are subjects that have been particularly productive in theoretical terms, even while other subjects may be of more immediate practical interest. The emphases of the book could have been different, but they could not have been very different while still being an introduction to central topics in S&TS.

An Introduction to Science and Technology Studies should provide an overview of the field for any interested reader not too familiar with S&TS's basic findings and ideas. The book might be used as a basis of a general upper-year undergraduate, or perhaps even graduate-level, course in S&TS. But it might also be used as part of a trajectory of more focused courses on, say, the social study of medicine, S&TS and the environment, reproductive technologies, science and the military, or science and public policy. Because anybody putting together such courses would have clear ideas on how those topics should be addressed – and certainly more knowledge on them than does the author of this book – these topics are not addressed here.

However the book is used, it would almost certainly be alongside a number of case studies, and probably alongside some of the many articles mentioned in the book. Though somebody approaching S&TS for the first time cannot read everything, the empirical examples here are not intended to replace rich detailed cases, but only to draw out a few salient features. Case studies are the bread and butter of S&TS. Almost all insights in the field grow out of them, and researchers and students still turn to articles based on cases to learn central ideas and to puzzle through problems. The empirical examples used in this book point to a number of canonical and useful studies, but of course there are many more that are not mentioned.

S.S.

Acknowledgments

A grant from the Social Sciences and Humanities Research Council of Canada facilitated some of my research on this book, as did a grant from the Queen's University Centre for Knowledge-based Enterprises. I learned an enormous amount from feedback on earlier versions by students in my "Science, Technology, and Society" course at Queen's; by Reijo Miettinen and his colleagues and students at the University of Helsinki, Petri Ylikoski, Juha Tuunainen, Erika Mattila, and Tarja Knuuttila; and especially by two anonymous reviewers for Blackwell Publishing. All authors should have such perceptive critics.

CHAPTER 1

The Prehistory of Science and Technology Studies

A View of Science

Let us start with a snapshot of a common view of science. It is a view that coincides more or less with where studies of science stood some 50 years ago, that still dominates popular understandings of science, and even serves as something like a mythic framework for scientists themselves. Snapshots always lack nuance, but they are useful nonetheless: this one includes a number of distinct elements and some healthy debates. It can, however, serve as an excellent foil for the discussions that follow. At the edges of this picture of science, and discussed in the next section, is technology, seen as simply the application of science.

In the common view, science is a formal activity that accumulates knowledge by directly confronting the natural world. That is, science makes progress because of its method, and because that method allows the natural world to play a role in the evaluation of theories. Science's method is a set of procedures and approaches that makes research systematic, and tends toward the discovery of truths. While the scientific method may be somewhat loose and large, and therefore may not level all differences, it creates a certain consistency: different scientists should perform an experiment similarly; scientists should be able to agree on important questions and considerations; and most importantly, different scientists considering the same evidence should accept and reject the same hypotheses. The result is that scientists can agree on truths about the natural world.

Exactly how science is a formal activity is open to some question. It is worth taking a closer look at some of the prominent views. Two important philosophical approaches within the study of science are *logical positivism*, initially associated with the Vienna Circle, and *falsificationism*, associated with Karl Popper. The Vienna Circle was a group of philosophers and scientists who met in the early 1930s. The project of the Vienna Circle was to develop a philosophical understanding of science that would allow for an expansion of the scientific worldview – particularly into the social sciences and into philosophy itself. The project was immensely successful, in the sense that positivism was widely absorbed by scientists and non-scientists interested in increasing the rigor of their work. Interesting problems in the view, however, meant that positivism became increasingly focused on issues within the philosophy of science, losing sight of the more general project with which the movement began (see Friedman 1999; Richardson 1998).

Logical positivists maintain that the meaning of a scientific theory (and anything else) is exhausted by considerations, logical and empirical, of what would verify or falsify it. A scientific theory, then, is in some sense a mere summary of possible observations, in a logically structured language. This is one way in which science can be seen as a formal activity: scientific theories are built up by the logical manipulation of observations (e.g. Ayer 1952 [1936]; Carnap 1952 [1928]), and scientific progress consists in increasing the number and range of potential observations that its theories indicate.

Theories develop through a method that transforms individual data points into general statements. The process of creating scientific theories is therefore an inductive one. As a result, positivists tried to develop a logic of science that would make solid the inductive process of moving from individual facts to general claims. For example, scientists might be seen as creating frameworks in which it is possible to unequivocally generalize from data (see box 1.1).

Positivism has immediate problems. First, if meanings are reduced to observations, there are many synonyms, theories or statements that look as though they should have very different meanings, but do not make different predictions. For example, Copernican astronomy was initially designed to duplicate the more successful predictions of the earlier Ptolemaic system; in terms of observations, then, the two systems were roughly equivalent, but they clearly meant very different things, since one put the Earth in the center of the universe, and the other had the Earth spinning around the Sun. Second, many apparently meaningful claims are not systematically related to observations, because theories are often too abstract to be immediately cashed out in terms of data. Abstraction should not render a theory meaningless. Despite these problems and others, the positivist view of meaning taps into deep intuitions, and cannot be entirely dismissed.

Box 1.1   The problem of induction

Among the asides inserted into the next few chapters are a number of versions of the "problem of induction." These are valuable background for a number of issues in Science and Technology Studies (S&TS). At least as stated here these are theoretical problems, that only occasionally become practical ones in scientific and technical contexts. While they could be paralyzing in principle, in practice they do not come up. One aspect of their importance, then, is in finding out how scientists and engineers contain these problems, and when they fail at that, how they deal with them.

The *problem of induction* arose with David Hume's general questions about evidence in the eighteenth century. Unlike classical skeptics, Hume was interested not in challenging particular patterns of argument, but in showing the fallibility of arguments from experience in general. In the sense of Hume's problem, induction is the extension of data to cover new cases. To take a standard example, "the sun rises every twenty-four hours" is a claim supposedly established by induction over many instances, as each passing day has added another data point to the overwhelming evidence for it. Inductive arguments take n cases, and extend the pattern to the $n+1st$. But, says Hume, why should we believe this pattern? Could the $n+1st$ case be different, no matter how large n is? It does no good to appeal to the regularity of nature, because the regularity of nature is one of the things at issue. And as Ludwig Wittgenstein (1958) and Nelson Goodman (1983 [1954]) show, nature could be perfectly regular and we would still have a problem of induction. This is because our ideas of what it means for the $n+1st$ case to be the same as the first n cases are not the only possible ones; *sameness* is not a fully defined concept.

It is intuitively obvious that the problem of induction is insoluble. It is more difficult to explain why, but Karl Popper, the political philosopher and philosopher of science, makes a straightforward case that it is. The problem is insoluble, according to him, because there is no principle of induction that is true. That is, there is no way of assuredly going from a finite number of cases to a true general statement about all the relevant cases. To see this, we need only look at examples. "The sun rises every twenty-four hours" is false, says Popper, as formulated and normally understood, because in Polar regions, there are days in the year when the sun never rises, and days in the year when it never sets. Even cases taken as examples of straightforward and solid inductive inferences can be shown to be wrong, so why should we be at all confident of more complex cases?

Even if one does not believe positivism's central ideas, many people are attracted to the strict relationship that it posits between theories and observations. Even if theories are not mere summaries of observations, they should be absolutely supported by them. The justification we have for believing a scientific theory is based on that theory's solid connection with data. Another view, then, that is more loosely positivist, is that one can by purely logical means make predictions of observations from scientific theories, and that the best theories are ones that make all the right predictions. This view is perhaps best articulated as *falsificationism*, a position developed by (Sir) Karl Popper (e.g. 1963), a philosopher who was once on the edges of the Vienna Circle.

For Popper, the key task of philosophy of science is to provide a demarcation criterion, a rule that would allow a line to be drawn between science and non-science. This he finds in a simple idea: genuine scientific theories are falsifiable, making predictions that are open to question. The scientific attitude demands that if a theory's prediction is falsified the theory itself is to be treated as false. Pseudo-sciences, among which Popper includes Marxism and Freudianism, are insulated from criticism, able to explain and incorporate any fact. They do not make any firm predictions, but are capable of explaining or explaining away anything that comes up.

This is a second way in which science might be seen as a formal activity. According to Popper, scientific theories are imaginative creations, and there is no method for creating them. As such, they are free-floating, their meaning not tied to observations as for the positivists. However, there *is* a strict method for evaluating them. Any theory that fails to make risky predictions is ruled unscientific, and any theory that makes failed predictions is ruled false. A theory that makes good predictions is provisionally accepted – until new evidence comes along. Popper's scientist is first and foremost skeptical, unwilling to accept anything as proven, and willing to throw anything away that runs afoul of the evidence. On this view, progress is probably best seen as the successive refinement and enlargement of theories to cover increasing data. While science may or may not reach the truth, the process of conjectures and refutations allows it to encompass increasing numbers of facts.

Like the central idea of positivism, falsificationism faces some immediate problems. Few scientific theories make hard predictions without adopting a whole host of extra assumptions (e.g. Putnam 1981); so most scientific theories are "unscientific." Also, when theories are used to make incorrect predictions, scientists often – and quite reasonably – look for reasons to explain away the predictions, rather than rejecting the theories. Nonetheless, there is something attractive about the idea that (potential) falsification is the key to solid scientific standing, and so falsificationism, like logical positivism, still has adherents today.

Box 1.2   The Duhem–Quine thesis

The *Duhem–Quine thesis* is the claim that a theory can never be conclusively tested in isolation: what is tested is an entire framework or a web of beliefs. This means that in principle any scientific hypothesis can be held in the face of apparently contrary evidence. Though neither of them put the issue quite this baldly, Pierre Duhem and W. V. O. Quine, writing in the beginning and middle of the twentieth century respectively, showed us why.

Let us imagine we have some theory that makes a number of predictions. How should we react if some of its predictions turn out to be false? The answer looks straightforward: the theory has been falsified, and should be abandoned. This is the basis of Popper's rationalist philosophy of science. But this answer is too easy, because theories never make predictions in a vacuum. Instead, they are *used*, along with many other resources, to make predictions. When a prediction is wrong, the culprit might be the theory. Alternatively, it might be the data that set the stage for the prediction, or additional hypotheses that were brought into play, or measuring equipment that was used to verify the prediction. The culprit might even lie entirely outside this constellation of resources: some unknown object or process that interferes with observations or affects the prediction.

To put the matter in Quine's terms, theories are parts of webs of belief. When a prediction is wrong, one of the beliefs no longer fits neatly into the web. To smooth things out – to maintain a consistent structure – one can adjust any number of the web's parts. Provided one is willing to redesign the web radically enough, any part of it can be maintained. One can even abandon rules of logic if one needs to!

Here is a simple example. When Newton's predictions of the path of the moon failed to match the data he had, he did not abandon his theory of gravity, his laws of motion, or any of the calculating devices he had employed. Instead, he assumed that there was something wrong with the observations, and he fudged his data. While fudging violates the scientific ethic, we can appreciate his impulse: in his view, the theory, the laws, and the mathematics were all stronger than the data! Later physicists agreed. The problem lay in the optical assumptions originally used in interpreting the data, and when those were changed Newton's theory made perfect predictions.

Does the Duhem–Quine thesis give us a problem of induction? It shows that any number of resources are used, implicitly and explicitly, to make a prediction, and that it is impossible to isolate only one of those resources as at fault when the prediction appears wrong. We might, therefore, see the Duhem–Quine thesis as posing a problem of deduction, not induction, because it shows that when dealing with the real world, many things can confound neat logical deductions.

For both positivism and falsificationism, the features of science that make it scientific are formal relations between theories and data, whether it is the rational construction of theoretical edifices on top of empirical data or the rational dismissal of theories on the basis of empirical data. There are analogous views about mathematics; indeed, we might see formalist pictures of science as depending on stereotypes of mathematics.

But there are other features of the popular snapshot of science. These formal relations between theories and data can be difficult to reconcile with an even more fundamental intuition about science: Whatever else it does, science progresses toward truth, and accumulates truths as it goes. We can call this intuition *realism*, the name that philosophers have given to the claim that many or most scientific theories are approximately true.

First, progress. One cannot but be struck by the increases in precision of scientific predictions, the increases in scope of scientific knowledge, and the increases in technical ability that stem from scientific progress. Even in so well-established a field as astronomy, calculations of the dates and times of astronomical events continue to become more precise. Sometimes this precision stems from better data, sometimes from better understandings of the causes of those events, and sometimes from connecting different pieces of knowledge. And occasionally, the increased precision allows for increased technical ability or new theoretical advances.

Second, truths. According to realist intuitions, there is no way to understand the overall increase in predictive power of science, and the technical ability that flows from that predictive power, except in terms of the increase of truth. That is, science can do more when its theories are better approximations of the truth, and when it has more approximately true theories. This means that science does not merely construct convenient theoretical descriptions of data, or merely discard falsified theories. When it constructs theories or other claims, those eventually approach the truth. When it discards falsified theories, it does so in favor of theories that better approach the truth.

However, realists are generally committed to something like formal relations between data and theories. Real progress has to be built on more or less systematic methods. Otherwise, there would only be occasional gains, stemming from chance or genius. If science accumulates truths, it does so on a rational basis, not through luck.

If we step away from issues of data, evidence, and truth, we see a social aspect to the standard picture of science. Scientists are distinguished by their even-handed attitude toward theories, data, and each other. Thus Robert Merton's *functionalist* view, discussed in chapter 3, dominated discussions of the sociology of science through the 1960s. Merton argued that science served a social function, providing certified knowledge. That function structures a number of norms of behavior: scientists adhering to those norms tend to contribute more to the scientific project than scientists not adhering to them. We have a picture of science as a relatively well-regulated activity, steadily adding to the store of knowledge. Merton does not describe a particular scientific epistemology, but a social structure that privileges epistemic concerns.

Merton does not claim that science is pursued by particularly "scientific" people. Rather, science's social structure rewards behavior that, in general, promotes the growth of knowledge; in principle it also penalizes behavior that retards the growth of knowledge. The need for progress, in the straightforward sense of the accumulation of knowledge, shapes the social structure of science. This position is held by a number of other thinkers, such as Popper (1963) and Michael Polanyi (1962), who both support an individualist, republican ideal of science, for its ability to progress. Interestingly, the position is even echoed by writers in the "Radical Science Movement," who, in the hopes of harnessing a fully objective science to socialist goals, are highly critical of science's economic and social roles.

Common to all of these views of science is the idea that standards or norms are the source of science's success and authority. For positivists, the key is that theories can be no more or less than the logical representation of data. For falsificationists, scientists are held to a standard on which they have to discard theories in the face of opposing data. For realists, good methods form the basis of scientific progress. For functionalists, the norms are the rules governing scientific behavior and attitudes. All of these standards or norms are attempts to define what it is to be scientific. They provide ideals that actual scientific episodes can live up to or not, standards to judge between good and bad science. Therefore, the view of science we have seen so far is not merely an abstraction from science, but is importantly a view of ideal science.

Box 1.3   Underdetermination

Scientists choose among competing hypotheses as offering the best account of some collection of data. This choice can never be logically conclusive, because for every explanation there are in principle an indefinitely large number of others that are exactly empirically equivalent. Theories are *underdetermined* by the empirical evidence. This is easy to see through an analogy.

Imagine that our set of data is the collection of points in the graph below. The hypothesis that we create to "explain" these data is some line of best fit. But what line of best fit? The graph on the right shows two competing lines that both fit the data perfectly.

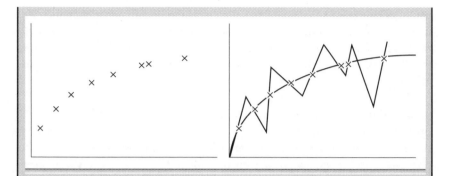

And clearly there are infinitely many more lines of perfect fit. We can do further testing and eliminate some, but there will always be infinitely many more. We can apply criteria like simplicity and elegance to eliminate some of them, but such criteria take us straight back to the first two problems of induction: how do we know that nature is simple and elegant, and why should we assume that our ideas of simplicity and elegance are the same as nature's?

When scientists choose the best theory, then, they choose the best theory from among those that have been seriously considered. There is little reason to believe that the best theory so far considered, out of the infinite numbers of empirically adequate explanations, will be the true one. In fact, if there are an infinite number of potential explanations, we could reasonably assign to each one a probability of zero.

The status of underdetermination has been hotly debated in philosophy of science. Because of the underdetermination argument, many philosophers (positivists and their intellectual descendants) argue that all scientific theories should be thought of as instruments for explaining and predicting, not as true or realistic representations (e.g. van Fraassen 1980). Realist philosophers, however, argue that there is no way of understanding the successes of science without accepting that in at least some circumstances evaluation of the evidence leads to approximately true theories (e.g. Boyd 1984; see box 6.2).

A View of Technology

Where is technology in all of this? Technology occupies a secondary role in this constellation of beliefs. This is for a simple reason: technology is often seen as the relatively straightforward application of science, in both popular and academic accounts. Technologists identify needs, problems, or opportunities, and creatively combine pieces of knowledge to address them. Technology combines the scientific method with a practically minded creativity. As such, the interesting questions about technology are about its effects: Does technology determine social relations? Is technology humanizing or dehumanizing? Does technology promote or inhibit freedom? These are important questions, but as they all take technology as a finished product they are normally pursued relatively divorced from considerations of the creation of particular technologies.

If technology is applied science then it is limited by the limits of scientific knowledge. On the common view, then, science plays a central role in determining the shape of technology. There is another form of determinism that often arises in discussions of technology, though one that has been more publicly controversial. A number of writers have argued that the state of technology is the most important cause of social structures, because technology enables most human action. People act in the context of available technology, and therefore people's relations among themselves can only be understood in the context of technology. While this sort of claim is often challenged – by people who insist on the priority of the social world over the material one – it has helped to focus debate almost exclusively on the *effects* of technology.

Lewis Mumford (1934, 1967) established an influential line of thinking about technology. According to Mumford, technology comes in two varieties. *Polytechnics* are "life-oriented," integrated with broad human needs and potentials. Polytechnics produce small-scale and versatile tools, useful for pursuing many human goals. *Monotechnics* produce "megamachines" that can increase power dramatically, but by regimenting and dehumanizing. A modern factory can produce extraordinary material goods, but only if workers are disciplined to participate in the working of the machine. This distinction continues to be a valuable resource for analysts and critics of technology (see, e.g. Franklin 1990; Winner 1986).

In his widely read essay "The Question concerning Technology" (1977 [1954]), Martin Heidegger develops a similar position. For Heidegger, distinctively modern technology is the application of science in the service of power; this is an objectifying process. In contrast to the craft tradition that produced individualized things, modern technology creates resources, objects made to be used. From the point of view of modern technology, the world consists of resources to be turned into new resources.

For both Mumford and Heidegger modern technology is shaped by its scientific rationality. Even the pragmatist philosopher John Dewey (e.g. 1929), who argues that science is simply theoretical technology, and that all rational thought is instrumental, sees technology (in the ordinary sense) as applied science. The view that technology is applied science tends toward a form of technological determinism. For example, Jacques Ellul (1964) defines *tech-*

nique as "the totality of methods rationally arrived at and having absolute efficiency (for a given stage of development)" (quoted in Mitcham 1994: 308). A society that has accepted modern technology finds itself going down a path of increasing efficiency, allowing technique to enter more and more domains.

Thus the view that a formal relation between theories and data lies at the core of science informs not only our picture of science, but of technology. Challenges to that view form the origins of S&TS.

A Preview of S&TS

S&TS starts from an assumption that science and technology are thoroughly social activities. They are social in that scientists and engineers are always members of communities, trained into those communities and necessarily working within them. Communities, among other things, set standards for inquiry and evaluate knowledge claims; there is no abstract and logical scientific method apart from the actions of scientists and engineers. In addition, science and technology are arenas in which rhetorical work is crucial, because scientists and engineers are always in the position of having to convince their peers and others of the value of their favorite ideas and plans – they are constantly engaged in struggles to gain resources and to promote their views. The actors in science and technology are also not mere logical operators, but instead have investments in skills, prestige, knowledge, and specific theories and practices. Thus ideology and values of many different types are important components of research. Even conflicts in a wider society may be mirrored by and connected to conflicts within science and technology; for example, splits along gender, race, class, and national lines can occur both within science and in the relations between scientists and non-scientists.

S&TS takes a variety of anti-essentialist positions with respect to science and technology. Neither science nor technology is a natural kind, having simple properties that defines it once and for all. The sources of knowledge and artifacts are complex and various: there is no scientific method to translate nature into knowledge, and no technological method to translate knowledge into artifacts. In addition, the interpretations of knowledge and artifacts are complex and various: claims, theories, facts, and objects may have very different meanings to different audiences.

For S&TS, then, science and technology are active processes, and should be studied as such. The field investigates how scientific knowledge and technological artifacts are *constructed*. Knowledge and artifacts are human products, and marked by the circumstances of their production. In their most crude forms, claims about the social construction of knowledge leave no role for the material world to play in the making of knowledge about it.

Almost all work in S&TS is more subtle than that, exploring instead the ways in which the material world is used by researchers in the production of knowledge. S&TS pays attention to the ways in which scientists and engineers attempt to construct stable structures and networks, often drawing together into one account the variety of resources used in making those structures and networks. Scientists and engineers *use* the material world in their work; it is not merely translated into knowledge and objects by a mechanical process.

Clearly, S&TS tends to reject many of the elements of the common view of science. How it gets to that point, and the ways in which it does so are the topics of the rest of this book.

CHAPTER 2

The Kuhnian Revolution

Thomas Kuhn is the best-known recent opponent of formalist pictures of science. His *The Structure of Scientific Revolutions* (1970a, first published in 1962) challenged the dominant popular and philosophical pictures of the history of science. In place of the formalist view with its normative stance, Kuhn substituted a focus on the activities around scientific research: at least officially, he insisted that science was merely what scientists do. Kuhn's picture does include large historical patterns, but in place of steady progress he substituted periods of normal science punctuated by revolutions. These innovations were in part an ingenious reworking of portions of the standard pictures of science, informed by rationalist emphases on the power of ideas, by positivist views on the nature and meaning of theories, and by Ludwig Wittgenstein's ideas about observation and about forms of life. The result was quite novel, and had an enormous impact.

According to Kuhn, *normal science* is what gets done when members of a field substantially share a recognition of key past achievements in their field, beliefs about which theories are right, an understanding of the important problems of the field, and methods for solving those problems. In Kuhn's flexible terminology, scientists doing normal science share a *paradigm*. The term, originally referring to a grammatical model or pattern, draws particular attention to an achievement that serves as an example for others to follow. Kuhn also assumes that those achievements provide theoretical and methodological tools for further research. Once they were established, Newton's mechanics, Lavoisier's chemistry, and Mendel's genetics each structured research in their respective fields, providing theoretical frameworks for and models of successful research.

Although it is tempting to see it as a period of stasis, normal science is better viewed as a period in which research is well structured. The theoretical side of a paradigm serves as a *world view*, providing categories and frameworks into which to slot phenomena. The practical side of a paradigm serves as a *form of life*, providing patterns of behavior or frameworks for action. But within paradigms research goes on, often with tremendous creativity – it is merely that creativity is embedded in a firm contextual backdrop.

Kuhn talks about normal science as *puzzle-solving*, because problems are to be solved within the terms of the paradigm: failure to solve such a problem typically reflects badly on the researcher, rather than on the theories or methods of the paradigm. With respect to a paradigm, an unsolved problem is simply an anomaly, fodder for future researchers. In periods of normal science the paradigm is not open to serious question. This is because the natural sciences, on Kuhn's view, are particularly successful at socializing practitioners. Science students are taught from textbooks that present standardized views of fields and their histories; they have lengthy periods of training and apprenticeship; and during their training they are generally asked to solve well-understood and well-structured problems.

Nothing good lasts for ever, and that includes normal science. Because paradigms can only ever be partial representations and partial ways of dealing with a subject matter, anomalies accumulate, and may eventually start to take on the character of real problems, rather than mere puzzles. Real problems cause discomfort and unease with the terms of the paradigm, and allow scientists to consider changes and alternatives to the framework; Kuhn terms this a period of *crisis*. If an alternative is created that solves some of the central unsolved problems, then some scientists, particularly younger scientists who have not yet been fully indoctrinated into the beliefs and way of life of the older paradigm, will adopt the alternative. Eventually, as older and conservative scientists become marginalized, a robust alternative may become a paradigm itself, structuring a new period of normal science.

Box 2.1   The modernity of science

Many commentators on science have felt that it is a particularly modern institution. By this they generally mean that it is exceptionally rational, or exceptionally free of local contexts. While science's exceptionality in either of these senses is contentious, there is a straightforward sense in which science is, and always has been, modern. As Derek de Solla Price (1986 [1963]) has pointed out, science has grown rapidly over the past 300 years. In fact, by any of a number of indicators, science's growth has been steadily exponential. Science's share of the US Gross National Product has doubled every 20 years. The cumulative number of scientific journals founded has doubled every 15 years, as has the membership in scientific institutes, and the number of people with scientific or technical degrees. The numbers of articles in many sub-fields have doubled every 10 years. These patterns cannot continue indefinitely – and in fact have not continued since Price did his analysis.

A feature of this extremely rapid growth is that between 80 and 90 percent of all the scientists who have ever lived are alive now. For a senior scientist, between 80 and 90 percent of all the scientific articles ever written were written during his or her lifetime. For working scientists the distant pasts of their fields are almost entirely irrelevant to their current research, because the past is buried under masses of more recent accomplishments. And citation patterns show, as one would expect, that older research is considered less relevant than more recent research, perhaps having been superseded or simply left aside. For Price, a "research front" in a field at some time can be represented by the network of articles that are frequently cited. The front continually picks up new articles and drops old ones, as it establishes new problems, techniques, and solutions. Whether or not there are paradigms as Kuhn sees them, science pays most attention to current work, and little to its past. Science is modern in the sense of having a present-centered outlook, leaving its past to historians.

According to Kuhn, it is in periods of normal science that we can most easily talk about progress, because problems are solved in a cumulative way, because knowledge is made more precise, and because scientists have little difficulty recognizing each other's achievements. Revolutions, however, are not progressive, because they both build and destroy. Some or all of the research structured by the pre-revolutionary paradigm will fail to make sense under the new regime; in fact Kuhn even claims that theories belonging to different paradigms are *incommensurable* – lacking a common measure – because people working in different paradigms see the world differently, and because the meanings of theoretical terms change with revolutions (a view derived in part from positivist notions of meaning). The non-progressiveness of revolutions and the incommensurability of paradigms are two closely related features of the Kuhnian account that have caused many commentators the most difficulty.

If Kuhn is right, science does not straightforwardly accumulate knowledge, but instead moves from one more-or-less adequate paradigm to another. This is the most radical implication found in *The Structure of Scientific Revolutions*: Science does not track the truth, but creates different partial views that can be considered to contain truth only by people who hold those views!

Box 2.2   Foundationalism

Foundationalism is the thesis that knowledge can be traced back to firm foundations. Typically those foundations are seen as a combination of sensory impressions and rational principles, which then support an edifice of higher-order beliefs. The central metaphor of foundationalism, of a building firmly planted in the ground, is an attractive one. If we ask why we hold some belief, the reasons we give come in the form of another set of beliefs. We can continue asking why we hold *these* beliefs, and so on. Like bricks, each belief is supported by more beneath it (there is a problem here of the nature of the mortar that holds the bricks together, but we will ignore that). Clearly, the wall of bricks cannot continue downward for ever; we do not support our knowledge with an infinite chain of beliefs. But what lies at the foundation?

The most plausible candidates for empirical foundations are sense experiences. But how can these ever be combined to support the complex generalizations that form our knowledge? We might think of sense experiences, and especially their simplest components, such as points in a visual field, as like individual data points. Here we have the earlier problems of induction all over again: as we have seen, a finite collection of data points cannot determine which generalizations we should believe.

Worse, even beliefs about sense impressions are not perfectly secure. Much of the discussion around Kuhn's *Structure of Scientific Revolutions* (1970a [1962]) has focused on his claim that scientific revolutions change what scientists see (box 2.3), a clear challenge to foundationalism. Even if Kuhn's emphasis is wrong, it is clear that we often doubt what we see or hear, and reinterpret it in terms of what we know. And the problem becomes more obvious if we imagine the foundations to be already-ordered collections of sense impressions – as the discussion of the Duhem–Quine thesis (box 1.2) shows.

On the one hand, then, we cannot locate plausible foundations for the many complex generalizations that form our knowledge. On the other hand, nothing that might count as a foundation is perfectly secure. Our best option is to abandon, then, the metaphor of solid foundations on which our knowledge sits.

Kuhn's claim that theories within paradigms are incommensurable has a number of different roots. One of those roots lies in the positivist picture of the meaning. Kuhn adopts the idea that the meanings of theoretical terms depend upon the constellation of claims in which they are embedded. Thus a change of paradigms should result in widespread changes in the meanings of key terms. But if this is true, then none of the key terms from one paradigm will map neatly on to those of another, preventing a common measure.

In *The Structure of Scientific Revolutions*, Kuhn takes the notion of indoctrination quite seriously, going so far as to claim that paradigms even shape observations. People working within different paradigms see things differently. Borrowing from the work of N.R. Hanson (1958), Kuhn argues there is no such thing, at least in normal circumstances, as raw observation. Instead, observation comes interpreted: we do not see dots and lines in our visual fields, but instead see more or less recognizable objects and patterns. Thus observation is guided by concepts and ideas. This claim has become known as the *theory-dependence of observation* (see box 2.3). The theory-dependence of observation is easily linked to Kuhn's historical picture, because during revolutions people stop seeing one way, and start seeing another way, guided by the new paradigm.

Box 2.3   The theory-dependence of observation

Do the beliefs that people hold shape what they observe? Psychologists have long studied this question, showing how people's interpretations of images are affected by what they expect those images to show. Norwood Russell Hanson and Thomas Kuhn took the psychological results to be important for understanding how science works. Scientific observations, they claim, are theory-dependent.

For the most part, philosophers, psychologists, and cognitive scientists agree that observations can be shaped by what people believe. There are substantial disagreements, though, about how important this is for understanding science. For example, a prominent debate about visual illusions and the extent to which the background beliefs that make them illusions are plastic (e.g. Churchland 1988; Fodor; 1988) has been sidelined by a broader interpretation of "observation." Many scientific data are collected by machine, and then are organized by scientists to display phenomena (Bogen and Woodward 1992). If that organization amounts to observation, then it is straightforward that observation is theory-dependent, seen from historical cases. But theory-dependence is broader than that: scientists attend to objects and processes that background beliefs suggest are worth looking at, they design experiments around theoretically inspired questions, they remember relevant and communicate relevant information, where relevance depends on theoretical views (Brewer and Lambert 2001). Whether or not observation, narrowly construed, is theory-dependent, the rest of science clearly is.

Finally, one of the roots of Kuhn's claims about incommensurability is his experience as an historian that it is difficult to make sense of past scientists' problems, concepts, and methods. Past research can be opaque, and aspects of it can seem bizarre, as seen in, for example, the nineteenth-century Naturphilosophie movement (e.g. Jardine 1991). It might even be said that if people find it too easy to assimilate very old research to present sensibilities, they are probably doing some interpretive violence to that research – Isaac Newton's physics looks strikingly modern when it is re-written for today's textbooks, but looks much less so in its originally published form, and even less so when the connections between it and Newton's religious and alchemical research are drawn (e.g. Dobbs and Jacob 1995). Kuhn says: "In a sense that I am unable to explicate further, the proponents of competing paradigms practice their trades in different worlds" (1970a, 150).

The semantic justification for incommensurability has attracted a considerable amount of attention, mostly negative. Meanings of theoretical terms do change, but they probably do not change so much and so systematically that claims in which they are used cannot typically be compared. Most of the philosophers, linguists, and others who have studied this issue have come to the conclusion that claims for semantic incommensurability cannot be sustained, or even that it is impossible (Davidson 1974) to make sense of such radical change in meaning (see Bird 2000, ch. 5, for an overview).

This leaves the historical justification for incommensurability. That problems, concepts, and methods change is uncontroversial. But the difficulties that these create for interpreting past episodes in science can be overcome – the very fact that historical research can challenge present-centered interpretations shows the limits of incommensurability.

Claims of radical incommensurability appear to fail. In fact, Kuhn quickly distanced himself from the strongest readings of his claims. Already by 1965 he insisted that he meant by "incommensurability" only "incomplete communication" or "difficulty of translation," sometimes leading to "communication breakdown" (Kuhn 1970b). On these more modest readings of incommensurability, though, incommensurability is an important phenomenon: even when dealing with the same subject matter, scientists (among others) can fail to communicate.

If there is no radical incommensurability, then there is no radical division between paradigms, either. Paradigms must be linked by enough continuity of concepts and practices to allow communication. This may even be a methodological or theoretical point: complete ruptures in ideas or practices are inexplicable (Barnes 1982). When historians want to explain an innovation, they do so in terms of a reworking of available resources. Every new idea, practice, and object has its sources; to assume otherwise is to invoke something akin to magic. Thus many historians of science have challenged Kuhn's paradigms by showing the continuity from one putative paradigm to the next.

For example, instruments, theories, and experiments change at different times. In a detailed study of particle detectors in physics, Peter Galison (1997) shows that new detectors are initially used for the same types of experiments and observations as their immediate predecessors had been, and fit into the same theoretical contexts. Similarly, when theories change, there is no immediate change in either experiments or instruments. Discontinuity in one realm, then, is at least generally bounded by continuity in others. Science gains strength, an *ad hoc* unity, from the fact that its key components rarely change together. Like Wittgenstein's thread – "the strength of the thread does not reside in the fact that some one fibre runs through its whole length, but in the overlapping of many fibres" – science maintains stability through change by being disunified. If Galison is right then the image of complete breaks between periods is wrong.

It is worth mentioning a feature of Kuhn's method which, while not original to him, is important to S&TS. One of the targets of *The Structure of Scientific Revolutions* is what is known (since Butterfield 1931) as *Whig history*, history that attempts to construct the past as a series of steps toward (and occasionally away from) present views. Especially in the history of science there is a temptation to see the past through the lens of the present, to see moves in the direction of what we now believe to be the truth as more rational, more natural, and less needing of explanation than opposition to what we now believe. But since events must follow their causes, a sequence of events in the history of science cannot be explained simply by the fact that they represent progress toward the truth. Whig history is one of the common buttresses of too-simple progressivism in the history of science, and its removal makes room for more jerky pictures.

The Structure of Scientific Revolutions had an immediate impact. The word "paradigm," referring to a way things are done or seen, came into common usage largely because of Kuhn. Even in the above short description of the book it is clear that it represents a profound challenge to the mid-century constellation of beliefs about science, even though it is partly a rearrangement of them! (Kuhn had less immediate impact on ideas about technology.)

Against the image of science with which we started, *The Structure of Scientific Revolutions* can be seen to indicate that scientific communities are most importantly organized around ideas and practices, not around ideals of behavior. And, they are organized from the bottom up, not, as functionalism would have it, to serve an overarching goal. Against positivism Kuhn argued that changes in theories are not driven by data but by changes of vision. In fact, if world views are essentially theories then data are subordinate to theory, rather than the other way around. (There were elements of this view already present in early positivism, but that was not widely recognized in the 1960s and 1970s.) Against falsificationism Kuhn argued that anomalies are typically set aside, that only during revolutions are they used as a justification to reject a theory. And against all of these he argued that on the large scale the history of science could not neatly be seen as a story of progress, only change.

Because Kuhn's version of science violated almost everybody's ideas of the rationality and progress of science, *The Structure of Scientific Revolutions* was sometimes read as a claim that science is fundamentally irrational, that it describes science as "mob rule." In retrospect it is difficult to find much irrationalism there, and easy to see the book as fairly conservative – Steve Fuller (2000b) sees the book as not only intellectually conservative but politically conservative. More important, perhaps, is the widespread perception that by examining history Kuhn firmly refuted the standard view of science. Whether or not that is true, it started people thinking about science in very different terms. The success of the book created a space for thinking about the practices of science in local terms, rather than in terms of their contribution to progress, or their exemplification of ideals. Though few of Kuhn's specific ideas have survived S&TS intact, *The Structure of Scientific Revolutions* has profoundly affected subsequent thinking in the study of science and technology.

CHAPTER 3

Questioning Functionalism in the Sociology of Science

Structural-functionalism

Robert Merton's statement "The institutional goal of science is the extension of certified knowledge" (1973: 270) is the supporting idea behind his thinking on science. His structural-functionalist view assumes that society as a whole can be analyzed in terms of overarching institutions such as religion, government, and science. Each institution, when working well, fulfills a necessary function, contributing to the stability and flourishing of the society. To work well, these institutions must have the appropriate social structure. Merton treats science, therefore, as a roughly unified and singular institution, the function of which is to provide knowledge. The work of the sociologist is primarily to study how its social structure does and does not support its function. Merton is the most prominent of functionalist sociologists of science, and so his work is the main focus of this chapter, to the neglect of such sociologists as Joseph Ben-David (1991) and John Ziman (1984), and such sociologically minded philosophers as David Hull (1988).

The key to Merton's theory of the social structure of science lies in the ethos of science, the norms of behavior that guide appropriate scientific practice. Norms are institutional imperatives, in that rewards are given to community members who follow them, and sanctions are applied to those who violate them. Most important in this ethos are the four norms that Merton first described in 1942, in an article later entitled "The Normative Structure of Science": universalism, communism, disinterestedness, and organized skepticism.

Universalism requires that the criteria used to evaluate a scientific claim not depend upon the identity of the person making the claim: "race, nationality, religion, class, and personal qualities are irrelevant" (Merton 1973: 270). This should stem from the supposed impersonality of scientific laws; they are either true or false, regardless of their proponents and their provenance. How does the norm of universalism apply in practice? We might look to science's many peer review systems. For example, most scientific journals accept articles for publication based on evaluations by experts in the relevant field. Typically, those experts are not told the identity of the authors whose articles they are reviewing. Although not being told the author's name does not guarantee the author's anonymity – because in many fields a well-connected reviewer can guess the identity of an author from the content of the article – it supports universalism nonetheless, both in practice and as an ideal.

Communism states that scientific knowledge – the central product of science – is commonly owned. Originators of ideas can claim recognition for their creativity, but cannot dictate how or by whom those ideas are to be used. Results should be publicized, so that they can be used as widely as possible. This serves the ends of science, because it allows researchers access to many more findings than they could hope to create on their own. According to Merton, communism not only promotes the goals of science but reflects the fact that science is a social activity, that scientific achievements are cumulatively produced. Even individuals' scientific discoveries come as a result of much earlier research.

Disinterestedness is a form of integrity, demanding that scientists disengage their interests from their actions and judgments. They are expected to report results fully, no matter what theory those results support. Disinterestedness should rule out fraud, such as reporting fabricated data, because fraudulent behavior typically represents the intrusion of interests into scientific work; thus Merton believes that fraud is rare in science.

Organized skepticism is the tendency for the community to disbelieve new ideas until they have been well established. Organized skepticism operates at two levels. New claims are often greeted by arrays of public challenges. For example, a presentation at a conference may be followed by fierce questioning, even by an audience favorably disposed to its claims. In addition, scientists may privately reserve judgment on new claims, employing a methodological version of the norm. In the latter case, scientists internalize the institutional approach.

In addition to these "moral" norms there are "cognitive" norms concerning rules of evidence, the structure of theories, and so on. Because Merton drew a relatively sharp distinction between social and technical domains, cognitive norms are not a matter for his sociology of science to investigate. Merton's sociology does not make any substantial claims about the content of science.

Institutional norms have effects in combination with rewards and sanctions, and in contexts in which community members are socialized to respond to those rewards and sanctions. Rewards in the scientific community are almost entirely honorific. As Merton identifies them, the highest rewards come via eponymy: *Darwinian biology,* the *Copernican system, Planck's constant,* and *Halley's comet* all recognize large achievements. Other forms of honorific reward are prizes, such as the Nobel Prize, and historical recognition; the most ordinary form of scientific reward is citation of one's work by others, seen as an indication of influence. Sanctions are similarly in terms of recognition, the reputations of scientists who display deviant behavior suffering accordingly.

In the 1970s, the Mertonian picture of the ethos of science came under attack, on a variety of instructive grounds. Although there were many criticisms, probably the two most important questions asked were: (1) Is the actual conduct of science governed by Mertonian norms? To be effective, norms of behavior must become part of the culture and institutions of science. In addition, there must be sanctions that can be applied when scientists deviate from the norms; but there is little evidence for the existence of strong sanctions for violation of these norms. (2) Are these norms too flexible to perform any analytic or scientific work? These and other questions created a serious challenge to that view, a challenge that helped to push science and technology studies (S&TS) toward more local, action-oriented views.

Ethos and Ethics

Social norms establish not only an ethos of science but an ethics of science. Violations of norms are importantly ethical lapses. This aspect of Merton's picture has given rise to some interesting attempts to understand and define scientific misconduct, a topic of increasing public interest (Guston 1999a).

In general, Merton thinks that the public nature of science means that deviant behavior is relatively rare. At the same time, deviance is to be expected, a result of structural conflicts among norms. In particular, science's reward system is the payment for communal ownership of results: scientific discoveries are made public, but their discoverers are credited and recognized. However, the pressures of recognition can often create pressures to violate other norms. A disinterested attitude towards one's data, for example, may go out the window when recognition is importantly at stake, and this may create pressure to fudge results. Fraud and other forms of scientific misconduct occur because of the structures that advance knowledge, not despite them.

Harriet Zuckerman (1977, 1984) refines Merton's model. Questions of misconduct often run into a problem of differentiating between fraud and error. Fraud is reprehensible, while error is merely undesirable. Zuckerman argues that the difference between them is the difference between the violation of social and that of technical norms. Among other things, this has the interesting consequence that intention is not at issue, that social norms in science can be violated whether or not scientists intend to violate them. In addition, the violation of social norms is more serious if it has more effect on the progress of knowledge.

Such models continue to shape discussions of scientific misconduct. The National Academy of Science's primer on research ethics, *On Being a Scientist* (1995), is a widely circulated booklet containing discussions of different scenarios and principles. Ethical norms, more concrete and nuanced than Merton's, are presented as in the service of the advancement of knowledge. That is, the resolution of ethical problems in science typically turns on understanding how to best maintain the scientific enterprise. Functionalism about science, then, can translate more or less directly into ethical advice.

Box 3.1   Is fraud common?

There are enormous pressures on scientists to perform, and to establish careers. At the same time there are difficulties in replicating experiments, there is an elite system that allows some researchers to be relatively immune from scrutiny, and there is an unwillingness of the scientific community to level accusations of outright fraud (Broad and Wade 1982). For these reasons it is difficult to know just how common fraud is, but many people suspect that it might be common.

Certainly some people feel that scientific fraud is common enough to be a major concern. Because of its substantial role in funding scientific research, the US Congress has on several occasions held hearings to address fraud. Prominently, Congressman Albert Gore, Jr., held hearings in 1981 in response to a rash of allegations of fraud in biomedical research at prominent institutions, and Congressman John Dingell held a series of hearings starting in 1988, that featured prominently "the Baltimore case," the subject of a book-length study by Daniel Kevles (1998).

David Baltimore was a Nobel Prize-winning biologist who became tangled up in accusations against one of his co-authors on a 1986 publication. The events became "the Baltimore case" because he was the most prominent of the scientific actors, and because Baltimore persistently and sometimes pugnaciously defended the accused researcher, Thereza Imanishi-Kari. In 1985, Imanishi-Kari was an immunologist at the Massachusetts Institute of Technology (MIT), under pressure to do enough research and to publish enough to merit tenure. She collaborated with Baltimore and four other researchers on an experiment on DNA rearrangement, the results of which were published in the journal *Cell*. A post-doctoral researcher in Imanishi-Kari's laboratory, Margot O'Toole, was assigned some follow-up research, but was unable to repeat the original results. Furthermore, O'Toole felt that the published data were importantly not the same as the data contained in the laboratory notebooks.

After a falling-out between Imanishi-Kari and O'Toole, and also between Imanishi-Kari and a graduate student, Charles Maplethorpe, who had become involved in the issue, questions about fraud started working their way up through MIT. Despite being settled in Imanishi-Kari's favor at the university, the questions made their way to Washington when Maplethorpe alerted National Institutes of Health scientists Ned Feder and Walter W. Stewart about the controversy. Because of their involvement in an earlier case, Feder and Stewart had become magnets for, and were on their way to becoming advocates of, the study of scientific fraud; they brought the case to the attention of Congressman Dingell.

Once it entered the US Congress, the case became a much larger confrontation. Baltimore defended Imanishi-Kari and attacked the inquiry as a witch-hunt; a number of his scientific colleagues thought his tack unwise, because of the publicity he generated, and because he was increasingly seen as an interested party. Dingell found in Baltimore an opponent who was important enough to be worth taking down, and found in O'Toole a convincing witness. In the course of the hearings, Baltimore's conduct was made to look unprofessional, to the extent that he resigned his position as president of Rockefeller University.

Imanishi-Kari was eventually exonerated by the Office of Research Integrity, and Baltimore was seen as having taken a courageous stand. In his study of the case, Kevles makes it clear that he believes that this was the right outcome. For him, a small dispute with personal overtones got blown out of proportion by entering a public stage. This raises questions about the nature of any accusation of fraud. At the same time, though, Kevles's research reinforces reasoning that raises suspicions about the commonness of scientific fraud: the pressure to publish was substantial; the experiments were difficult to repeat; whether there had been fraud, or even substantial error, was open to interpretation; and the local scientific investigation was quick, though perhaps correctly quick, to find no evidence of fraud.

Is the Conduct of Science Governed by Mertonian Norms?

Are the norms of science constant through history and across science? A cursory look at different broad periods suggests that they are not constant, and consideration of different roles scientists can play – in industry, in the academy – shows that at the least norms will be interpreted differently by different actors (e.g. Zabusky and Barley 1997). Are they distinctive to science? Universalism, disinterestedness, and organized skepticism are at some level *professed* norms for many activities in many societies, and may not be statistically more common in science than elsewhere. Disinterestedness, for example, is a rewriting of a norm of rationality, but as a professed norm rationality is found nearly everywhere; people inside and outside of science believe that they generally act rationally. What evidence could show us that science is more rational than other activities?

As we saw in the last chapter, in *The Structure of Scientific Revolutions* Kuhn describes the work of normal science as governed by a paradigm. Focusing on the intellectual side of paradigms, normal science is governed by ideas that are specific to particular areas of research at particular times. For Kuhn, science is shaped by solidarities built around key ideas, not around behaviors. In Mertonian terms, such ideas are "cognitive norms," subsidiary to more general norms. Barry Barnes and R. G. A. Dolby (1970) suggest that cognitive norms are more important to scientists' work than are any general moral norms. For example, on the Kuhnian picture scientific education is authoritarian, militating against skepticism in favor of specific general beliefs.

This point can be seen in another criticism of Mertonian norms, put forward by Michael Mulkay (1969), using the example of the furor over the work of Immanuel Velikovsky. In his book *Worlds in Collision* Velikovsky argued that historical catastrophes, recorded in the Bible and elsewhere, were the result of a near-collision between Earth and a planet-sized object that broke off from Jupiter. The majority of mainstream scientists saw this as sensational pseudo-science. Mulkay uses the case to show one form of deviance from Mertonian norms in science:

> In February, 1950, severe criticisms of Velikovsky's work were published in *Science News Letter* by experts in the fields of astronomy, geology, archaeology, anthropology, and oriental studies. None of these critics had at that time seen *Worlds in Collision*, which was only just going into press. Those denunciations were founded upon popularized versions published, for example, in *Harper's*, *Reader's Digest* and *Collier's*. The author of one of these articles, the astronomer Harlow Shapley, had earlier refused to read the manuscript of Velikovsky's book because Velikovsky's "sensational claims" violated the laws of mechanics. Clearly the "laws of mechanics" here operate as norms, departure from which cannot be tolerated. As a consequence of Velikovsky's non-conformity to these norms Shapley and others felt justified in abrogating the rules of universalism and organized skepticism. They judged the man instead of his work.   (Mulkay 1969: 32–3)

Scientists violated Mertonian norms in the name of a higher one: claims should be consistent with well-established truths. One could argue that, even on Mertonian terms, violation of the norms in the name of truth makes perfect sense, since those norms are supposed to represent a social structure that aids the discovery of truths. Nonetheless, this type of case shows one way in which moral norms are subservient to cognitive norms. Kuhn would claim that this sort of example is ubiquitous.

So far, we have seen that Merton's norms may not be central to understanding the practice of science. But what if we looked at the practice of science and discovered that the opposites of those norms – secrecy, particularism, interestedness, and credulity – were common? Do scientific communities and their institutions sanction researchers who are, say, secretive about their work? There are, after all, obvious reasons to be secretive about one's work. If other researchers learn about one's ideas, methods, or results, they may be in a position to use that information to take the next steps in a program of research, and receive full credit for whatever comes of those steps. Given that science is highly competitive, and given that more and more science is linked to applications on which there are possible financial stakes, there are strong incentives to follow through on a research program before letting other researchers know about it. On the structural-functionalist picture, norms exist to counteract local interests such as recognition and monetary gain, so that the larger goal – the growth of knowledge – is served. Thus we should expect to see violations of norms sanctioned.

In a study of scientists working on the Apollo moon project, Ian Mitroff (1974) shows not only that scientists do not apply sanctions, but they can be seen to respect what he calls counter-norms, which are rough opposites of Mertonian ones. Scientists interviewed by Mitroff voiced approval of, for example, interested behavior (1974: 588): "Commitment, even extreme commitment such as bias, has a role to play in science and can serve science well." "Without commitment one wouldn't have the energy, the drive to press forward against extremely difficult odds." "The [emotionally] disinterested scientist is a myth. Even if there were such a being, he probably wouldn't be worth much as a scientist." Mitroff's subjects identified positive value in opposites to each of Merton's norms: Claims are judged by who makes them. Secrecy is valued because it allows scientists to follow through on research programs without worrying about other people doing the same work. Dogmatism allows people to build on others' results without worrying about foundations.

If there are both norms and counter-norms, then the analytical framework of norms does no work. The analytic framework of norms and counter-norms can justify anything, which means that it does not help to understand anything. Moreover, this is not just a methodological problem. When scientists act, norms and counter-norms can give them no guidance and cannot cause them to do anything: the reasons for or causes of actions must lie elsewhere.

Interpretations of Norms

Norms have to be interpreted. This represents a problem for the analyst, but also shows that norms may not do much work. Let us return to Mulkay's example of the Velikovsky case, which he himself later criticized to underscore difficulties in the interpretation of norms. The example was originally used to show that scientists violated norms when a higher norm was at stake. Mulkay later noticed that, depending on which parts of the context one attended to, the norms could be interpreted as having been violated or not.

> It could be argued that the kind of qualitative, documentary evidence used by Velikovsky had been shown time and time again to be totally unreliable as a basis for impersonal scientific analysis and that to treat this kind of pseudo-science seriously was to put the whole scientific enterprise in jeopardy. In this way scientists could argue that their response to Velikovsky was an expression of organized skepticism and an attempt to safeguard universalistic criteria of scientific adequacy.   (Mulkay 1980: 112)

The problem here is an instance of a more general problem about following rules (box 3.2), because different behaviors can be interpreted as following (or not following) the norms. This sort of reasoning can be extended to almost any violation of norms, which poses serious problems for any attempt to analyze science in terms of general norms. We can see almost any scientific episode as one of adherence to Mertonian norms, or as one of the violation of those norms. Thus they do little analytic work.

As in the case of counter-norms, the problem is not just a methodological one. If we as onlookers can interpret the actions of scientists as either in conformity to or in violation of the norms, so can the participants themselves. But that simply means that the norms do not constrain scientists. By creatively selecting contexts, any scientist can use the norms to justify almost any action. And if norms do not represent constraints, then they do no scientific work.

Box 3.2   Wittgenstein on rules

Many people working in S&TS have found the work of Ludwig Wittgenstein to be particularly valuable. In particular, Wittgenstein's discussion of rules and following rules is routinely invoked as foundational. Although it is complex, the central point can be seen in a short passage. Wittgenstein asks us to imagine a student who has been taught basic arithmetic. We ask this student to write down a series of numbers starting with zero, adding two each time (0, 2, 4, 6, 8, . . .).

> Now we get the pupil to continue . . . beyond 1000 – and he writes 1000, 1004, 1008, 1012. We say to him: "Look what you've done!" – He doesn't understand. We say: "You were meant to add *two*: look how you began the series!" – He answers: "Yes, isn't it right? I thought that was how I was *meant* to do it." – Or suppose he pointed to the series and said: "But I went on in the same way." – It would now be no use to say: "But can't you see . . .?" – and repeat the old examples and explanations. – In such a case we might say, perhaps: It comes natural to this person to understand our order with our explanations as *we* should understand the order: "Add 2 up to 1000, 4 up to 2000, 6 up to 3000, and so on." (Wittgenstein 1958, Paragraph 185)

Of course this student can be corrected, and can be taught to apply the rule as we would – there is coercion built into such education – but there is always the possibility of future divergences of opinion as to the meaning of the rule. In fact, Wittgenstein says, "no course of action could be determined by a rule, because every course of action can be made out to accord with a rule" (Paragraph 201). Rules do not contain the rules for their own applicability.

It is relatively easy to see how Wittgenstein's problem is an extension of Hume's problem of induction. A finite number of examples, with a finite amount of explanation, cannot constrain the next unexamined case. The problem of rule-following becomes a usefully different problem, though, in the context of actions, and not justobservations. It becomes valuable to understanding the practicesinvolved in science and technology.

There are competing interpretations of Wittgenstein's writing on this problem. Some take him as posing a skeptical problem, and giving a skeptical solution: people come to agreement about the meaning of rules because of prior socialization, and continuing social pressure (Kripke 1982). Others take him as giving an anti-skeptical solution after showing the absurdity of the skeptical position: hence we need to understand rules not as formulas standing apart from their application, but as constituted by their application (Baker and Hacker 1984). The same debate has arisen within S&TS (Bloor 1992; Lynch 1992a, 1992b). For our purposes here it is not crucial which of these positions is right, either about the interpretation of Wittgenstein or about rules. Instead, we might note that both sides agree that expressions of rules do not determine their applications.

Norms as Resources

Recognizing that norms can be interpreted flexibly suggests that we study not how norms act, but how norms are used. That is, in the course of explaining and criticizing actions, scientists invoke norms – such as the Mertonian norms, but in principle an indefinite number of others. For example, because of his refusal to accept the truth of quantum mechanics, Albert Einstein is often seen as becoming conservative as he grew older; being "conservative" clearly violates the norm of disinterestedness (Kaiser 1994). Einstein is so labeled in order to understand how the same person who changed accepted notions of space and time could later reject a theory because it challenged accepted notions of causality: otherwise how could the twentieth century's prime example of a scientific genius make such a mistake? Implicit in the label, however, is an assumption that Einstein was wrong to reject quantum mechanics, an assumption that quantum mechanics is obviously right, whatever the difficulties that some people have with it. Indeed, Werner Heisenberg, one of the participants in the debate, discounted Einstein's positions by claiming that they were produced by closed-minded dogmatism and old age. If we believe Heisenberg, we can safely ignore critics of quantum mechanics. How are norms serving as resources in this case? They are being used to help eliminate conflicting views: because Einstein's opposition to quantum mechanics was based in violation of norms of conduct, we do not have to pay much attention to his arguments.

Whether a theory stands or falls depends upon the strengths of the arguments put forward for and against it (it also depends upon the theory's usefulness, upon the strengths of the alternatives, and so on). However, it is rarely simple to evaluate important and real theories, and so complex arguments are crucial to science and to scientific beliefs. Norms of behavior can play a role, if they are used to diminish the importance of some arguments and increase the importance of others. Supporters of quantum mechanics are apt to see Einstein as a conservative in his later years. Opponents of quantum mechanics are apt to see him as maintaining a youthful skepticism throughout his life (Fine 1986). Norms are ideals and, like all ideals, they do not apply straightforwardly to concrete cases. People with different interests and different perspectives will apply norms differently.

Thus we are led from seeing norms as constraining actions to seeing norms as rhetorical resources. This is one of many parallel shifts of vision in S&TS, of which we will see more in later chapters. For the most part, these are changes from more structure-centered perspectives to more actor-centered perspectives. That is not to say that there is one simple theoretical maneuver that characterizes S&TS, but only that the field has found some shifts from structure- to actor-centered perspectives particularly valuable.

Boundary Work

Thomas Gieryn's study of "boundary work" is one approach to seeing norms as resources. Gieryn gives particular weight to the issue of epistemic authority, the authority to make claims that will be respected. When issues of epistemic authority arise, people attempt to draw boundaries: to have authority on any contentious issue requires that at least some other people do not have it. The study of boundary work is an anti-essentialist approach to understanding authority (Gieryn 1999). For example, some people might argue that science gets its epistemic authority from its rationality, its connection to nature, or its connection to technology or policy. Gieryn, however, would see those connections as products of boundary work: science is rational because it has acquired power to define the bounds of rationality; science is connected to nature because it has authority to determine what nature is; and scientists connect their work to the benefits of technology or the urgency of political action in particular situations when they are seeking epistemic authority. Yet those same connections are made carefully, to protect the authority of science, and are countered by boundary work aimed at protecting or expanding the authority of engineers and politicians (Jasanoff 1987).

From the case of Cyril Burt (box 3.3) we can see the interpretation of norms, in the service of the boundaries and cognitive authority of disciplines. Science can be portrayed as a flexible activity, and dubious practices can be seen as mere examples of practical necessity. Or, science can be portrayed as more rigid and dubious practices as deviant. Norms of scientific behavior are subject to flexible interpretation, interpretation that serves specific goals in specific contexts.

Boundary work is a concept with broad applicability. Norms are not the only resources that can be used to stabilize or destabilize boundaries. Organizations can help to further goals while maintaining the integrity of established boundaries (Guston 1999b; Moore 1996). Examples, people, methods, and qualifications are all used in the practical and never-ending work of charting boundaries. Textbooks, courses, and museum exhibits, for example, can establish maps of fields simply through the topics and examples that they represent (Gieryn 1996). In fact, little does not participate in some sort of boundary work, since every particular statement contributes to a picture of the space of allowable statements.

Box 3.3   Cyril Burt from hero to fraud

Thomas Gieryn and Anne Figert (1986) look at a controversy over Sir Cyril Burt (1883–1971), one of the most eminent psychologists of the twentieth century, knighted for his contributions to psychology and to public policy. Burt was best known for his strong data and arguments supporting hereditarianism (nature) over environmentalism (nurture) about intelligence. After Burt's death, opponents of hereditarianism pointed out that his findings were curiously consistent over the years. In 1976 in *The Times*, a medical journalist, Oliver Gillie, accused Burt of falsifying data, inventing studies and even co-workers. This public accusation of fraud, against one of the discipline's most noted figures, posed a potential challenge to the authority of psychology itself.

Early on, his supporters represented Burt as occasionally sloppy, but insisted that there was no evidence of fraud. Burt's work was difficult, they argued, and it was therefore understandable that he would have made some mistakes. No psychologist's work would be immune from criticism. In addition, Burt was an "impish" character, thus explaining his invention of colleagues. These responses construed Burt's work as scientific, but science as imperfect. That is, psychologists drew boundaries that accepted minor flaws in science, and thus allowed a flawed character to be one of their own.

In addition to denying or minimizing the accusations, responses by psychologists involved charging Gillie with acting inappropriately. By publishing his accusations in a newspaper, Gillie had subjected Burt's work to a trial not by his peers. The public nature of the IQ controversy raised questions about motives: were environmentalists trumping up or blowing up the accusations to discredit the strongest piece of evidence against them? Psychologists insisted that there be a *scientific* inquiry into the matter, and thus endorsed the ongoing research of one of their own, Leslie Hearnshaw, who was working on a biography of Burt. That biography ended up agreeing with the accusers but rescued psychology as a whole. It did so by banishing Burt, and recovering authority by assuming that "the truth will out" in science. Hearnshaw argued that Burt's fraud was the result of personal crisis, especially late in life, and was thus the result of his acting in a particularly unscientific manner. Most importantly, though, he argued that Burt was not a real scientist, but was rather an outsider who sometimes did good scientific work:

> Burt was primarily an applied psychologist. . . . The gifts which made Burt an effective applied psychologist, however, militated against his scientific work. Neither by temperament nor by training was he a scientist. He was overconfident, too much in a hurry, too eager for final results, too ready to adjust and paper-over, to be a good scientist. His work often had the appearance of science, but not always the substance. (Hearnshaw, quoted in Gieryn and Figert 1986: 80)

The Place of Norms in Science?

The failings of Merton's functionalist picture of science are instructive. Merton can be seen as asking what science needs to be like, as a social activity, in order for it to best provide certified knowledge. His four norms provide an elegant solution to that problem, and a plausible solution in that they are professed standards of scientific behavior.

Nonetheless, these norms do not seem to describe the behavior of scientists, unless the framework is interpreted very flexibly. But if it is interpreted flexibly then it ceases to do real analytic or explanatory work. Going a little deeper, critics have also challenged the idea that science is a unified institution organized around a single goal or even a set of goals: Does the idea of an institutional goal, for an entity as large and diffuse as science, even make sense? Could an institutional goal for science have any effect on the actions of individual scientists?

As a result of these arguments, critics suggest that science is better understood as the product of scientists acting to pursue their own goals. Merton's norms, then, are ideological resources, available to scientific actors for their own purposes. They serve, combined with formalist epistemologies, as something like an "organizational myth" of science (Fuchs 1993).

Still, we can ask how ideologies like Merton's norms affect science as a whole. It may be that their repeated invocation leads to their having real effects on the shape of scientific behavior. We might expect, for example, that the repeated demand for universalism will lead to some types of discrimination being unacceptable – shaping the ethics of science. Along with other values, Merton's norms may contribute to what Lorraine Daston (1995) calls the "moral economy" of science. While science as a whole may not have institutional goals, combined actions of individual scientists might shape science to look as though it has goals (Hull 1988). Even though boundaries, in this case boundaries of acceptable behavior, are constructed, they can have real effects.

CHAPTER 4

Stratification and Discrimination

This chapter takes up one of the central questions of the functionalist program, whether the divisions that the reward system of science creates are justified. The chapter thus represents a break, for a topic of wide interest, in the intellectual narrative set up so far. Readers wanting to follow that narrative might turn directly to chapter 5.

An Efficient Meritocracy or an Inefficient Old-boy Network?

In a study of the 55 Nobel laureates working in the United States in 1963, Harriet Zuckerman (1977) found that a full 34 of them had studied or collaborated with a total of 46 previous prize winners. Not only that, but those who had worked with Nobel laureates before they themselves had done their important research received their prizes at an average age of 44, compared with an average age of 53 for the others. Clearly scientists tend to form elite groupings. Are these groupings the tip of a merit-based iceberg, or are they artifacts of systems of prestige gone awry? Is the knowledge for which elites are recognized intrinsically and objectively valuable, or does it become so because of its association with the elites?

The Renaissance thinker Francis Bacon thought that the inductive method he had set out for science would level the differences between intellects and allow science to be industrialized. Three hundred years later, the Spanish philosopher José Ortega y Gasset claimed that "science has progressed thanks in great part to the work of men astoundingly mediocre, and even less than mediocre." Despite such pronouncements, a small number of authors publish many papers and a great number publish very few. According to Price (1986), roughly 10 percent of all scientific authors produce 50 percent of all scientific papers. Similarly, a small number of articles are cited many times and a great number are cited very few times. Jonathan and

Stephen Cole (1973) estimate that 80 percent of citations are to 20 percent of papers. If citations represent influences then these figures suggest that a small number of publications are quite influential, and the vast majority make only a small contribution to future advances. To use citations as measures of influence can be misleading (box 4.1). However, these figures are so striking that even if citations are crude measures of influence, Bacon and Ortega are wrong about the structure of science.

It is useful to contrast two possible policy implications of such figures, to see some of their practical importance. One obvious possible conclusion, drawn by the Coles, is that scientific funding could be allocated much more efficiently, being placed in the hands of people more likely to write influential articles. Since science is already highly stratified, funding could be made to better reflect that stratification. Most of the 80 percent less-cited papers need not be written. In fact, they may be entirely replaceable, thus reducing the value of the already-small contribution they make: the Coles argue that, given the number of multiple discoveries in the history of science, discoveries happen regularly.

This policy recommendation is based on an assumption that citations recognize something like intrinsic value: if 80 percent of citations are to 20 percent of the articles, it is because that 20 percent is intrinsically more valuable. Stephen Turner and Daryl Chubin (1976) point out that by denying that assumption one could draw very different policy lessons from the same citation data. The data could be taken to suggest that science uses its resources poorly, and that the majority of perfectly good articles are ignored. Turner and Chubin's alternative policy recommendations would be aimed at leveling playing fields, reducing the effects of prestige and of old-boy networks. This disagreement turns on the question of how accurate science's system of recognition is, a question that is difficult to answer.

One might also raise very different questions about where science places value. Both the Coles and Turner and Chubin seem to assume that scientific publications are the only significant locus of scientific value. They share a perspective that sees science as essentially a producer of pieces of knowledge. However, it might be argued that scientific skills are a central value – for example, Chandra Mukerji (1989) claims that oceanography is supported by the state because it produces a pool of skilled oceanographers, who can be called upon to perform particular services. Or, it might be argued that, particularly in today's climate, scientific skills and knowledge are primarily in the service of technological change. While neither of these alternatives looks plausible as a universal statement about science, they show some of the difficulties with a singular focus on publications.

Box 4.1   Do citations tell us about influence?

A number of sociologists have attempted to refine citation analysis as a tool to evaluate the importance of particular publications, among other things. In a review of this work, David Edge (1979) challenges the idea that citation analysis is a useful tool for studying scientific communication.

A core assumption of citation analysis is that citations represent influence. A citation is supposed to be a reference to a publication that provides important background information. This is an idealization. As has long been recognized (e.g. Medawar 1963), scientific publications are often misleading about their own histories. An article stemming from an experiment looks as though it recounts the process of experimentation, but is actually a rational reconstruction of an experiment. It is written to fit into universal categories, and so, not only are details cleaned up and idiosyncrasies left out, but temporal order may be changed to create the right narrative. The scientific paper is typically an argument. Its citations, therefore, are vehicles for furthering its argument, not necessarily records of influences (Gilbert 1977). Citations are biased toward the types of references that are useful in addressing intended audiences. Meanwhile, informal communication and information that does not need to be cited – or is even best not cited – is all left out of citations. Even when citations are not intended to further arguments, they may serve other purposes, even several purposes simultaneously (see Case and Higgins 2000). Citation analysis is therefore a poor tool for studying communication and influence (see MacRoberts and MacRoberts 1996).

Contributions to Productivity

As a review by Mary Frank Fox (1983) shows, no obvious variable has an overwhelming effect on scientists' publication productivity (generally measured in terms of number of publications). However, on the basis of that survey one can build up a picture of the most productive scientist as someone who: is driven, has a strong ego, has a history of acting autonomously, plays with ideas, tolerates ambiguity, is at home with abstractions, is detached from non-scientific social relations, reserves mornings for writing, works on routinized problems, is relatively young, went to a prestigious graduate school, had a prestigious mentor, is currently at a prestigious place of employment, has considerable freedom to choose their own problems, has a history of success, and has had previous work cited (see Fox 1983).

Many of the items in this ungainly list are uncontroversial, but for somebody seeking to understand scientific processes they are not particularly helpful. Peak productive years for scientists tend to center around their mid-forties, but it is unclear why, or what could be done to increase the productivity of older scientists.

Scientists working at prestigious locations have strikingly higher productivity, and institutional prestige is a strong predictor of productivity (e.g. Allison and Long 1990; Long and McGinnis 1981). Do prestigious university departments and independent laboratories select researchers who are more productive? The evidence suggests that differences in productivity emerge *after* people have been hired. Do employers with more prestige select scientists who are going to be more productive? It is difficult to imagine how they would be able to as consistently as they do, especially since the academic job market has long been tight, and the differences between candidates who are hired at more and less prestigious institutions could be expected to be very slight. So, what is it about prestige that contributes to productivity? There are a number of possibilities. Facilities might be better at prestigious locations. Prestigious environments might provide more intellectual stimulation than do non-prestigious environments. Prestigious environments might also be ones in which the internal pressure and rewards for productivity are greatest. Or, prestigious environments might provide more motivation by providing more visibility, and therefore more rewards for researchers – if this is right then prestige by itself makes people's research valuable!

The *cumulative advantage* hypothesis ties many of the above variables together. Robert Merton (1973, ch. 20) coined the term *Matthew Effect* after the line from the Gospel according to St. Matthew (13, verse 12): "For whosoever hath, to him shall be given, and he shall have more abundance: but whosoever hath not, from him shall be taken away even that he hath." In science, success breeds success. People with some of the psychological traits and work habits in Fox's list above are more likely to have a famous mentor, and to study at a prestigious graduate school. Once there, they are more likely to gain employment in a prestigious department. People in better departments may have access to better facilities and intellectual stimulation. Perhaps more importantly, they are more visible and therefore more likely to be cited. Within the social structure of science, citation is a reward. Citation appears to be an effective reward, as authors whose work is cited tend to continue publishing (Lightfield 1971) – in Randall Collins's (1998) terms, citations are sources of emotional energy.

The cumulative advantage hypothesis is attractive not merely because it ties together so many insights. It is also attractive because it straddles a fine line between seeing science's reward structure as merit-based and seeing it as elitist or idiosyncratic. Rewards for small differences, which may or may not be differences in merit, can end up having large effects, including effects on productivity.

Discrimination

Women and some minority groups are progressively more poorly represented the more elite the grouping of scientists. Although science is no longer a monastic order (Noble 1992), in the United States in 1995, 45% of bachelor's degrees in the social sciences, natural sciences, and engineering went to women, but only 31% of doctorates went to women (National Science Foundation 1998). These figures obscure large differences among fields, however: for example, women earned 38% of doctoral degrees in the biological sciences but only 12% in engineering. The figures for women employed in the sciences are much more stark; for example, in the United States in 1995 only 16% of senior faculty in these areas were women, heavily concentrated in the life and social sciences. Such figures represent a marked improvement from 20 years earlier (emphasized, for example, by Long 2001), but they still indicate discrimination within science and engineering (Schiebinger 1999).

In addition to varying from field to field, women's participation in science and engineering varies quite substantially from country to country. Although nowhere is there equity, compared to the United States women are more poorly represented in the natural sciences in such places as the United Kingdom, Brazil, and Italy, but better represented in Turkey, Mexico, and Portugal (Etzkowitz, Kemelgor, and Uzzi 2000). In Turkey the figures are striking enough that one author has asked not "why so few?" but "why so many?" (Oncu, cited in Etzkowitz et al. 2000: 205) – the answer probably lies in a combination of a class structure that outweighs the gender structure and the relative marginality of science.

Natalie Angier uses a common metaphor, the pipeline, to describe the problem of women in science and engineering. Women flow through "a pipe with leaks at every joint along its span, a pipe that begins with a high-pressure surge of young women at the source – a roiling Amazon of smart graduate students – and ends at the spigot with a trickle of women prominent enough to be deans or department heads at major universities or to win such honours as membership in the National Academy of Science or even, heaven forfend, the Nobel Prize" (quoted in Etzkowitz et al. 2000: 5). The problem is typically seen as one of leakage: women are leaking out of the pipeline all the way along. However, the pipeline metaphor only makes sense if we think in terms of not only leakage but blockage and filters (Etzkowitz et al. 2000).

Similarly, Sara Delamont (1989) describes the problem of equity for women in terms of three problems in women's careers, "getting in," "staying on," and "getting on" (see also Glover 2000). If science is to move toward equity, women need to enter the pipeline, not leak out, and keep their careers moving forward.

Getting in, Staying in, and Getting on

Margaret Rossiter's (1982, 1995) acclaimed historical studies of women scientists in the United States present a fascinating set of pictures of the struggles of women to work within and change the cultures and institutions of science. For Rossiter, the overarching problem that women have faced in science has been that feminization is linked to a lowering of status and masculinization to an increase in status. When, in the 1860s and 1870s, women increasingly attempted to enter scientific fields, men in those fields perceived that their work was losing its masculine status (Rossiter 1982, ch. 4). As a result, they forced women into marginal positions in which they would be "invisible." In the twentieth century women found that they could enter the scientific world much more easily if they entered new fields, such as home economics and nutritional science. There they could do sociology, economics, biology, and chemistry; however, the more women there were in new fields, the more poorly they were regarded.

Similarly, women's colleges in the US had traditionally provided safe environments for women scientists, in which they could work and establish their credentials. However, during the expansion and rise of colleges and universities after the Second World War women's colleges and less prestigious universities followed the lead of more prestigious universities in hiring more men and in keeping women invisible. Thus the "golden age" of science in the US after the Second World War was a "dark age" for women in science (Rossiter 1995: xv).

Rossiter argues that there was a qualitative change to women's place in US science only in the 1960s. At that point women developed some rhetorically powerful language for describing the discrimination that they faced. The National Organization for Women and activists like Betty Friedan articulated issues of women's rights in terms that resonated with US ideals. Alice Rossi's 1965 article in *Science*, entitled "Women in Science: Why So Few?" was a key document in the articulation of the problems women faced in professional life, and made clear that science was no different from other areas. At the same time, and obviously related, there began a legal revolution that resulted in new rights for women, in both education and employment. While the problem of the status of feminized fields persists, the new anti-discrimination laws and affirmative action programs have been useful resources in women's struggles to "get in" to the worlds of professional science and engineering.

Getting into science is still not a given for women. Science has a masculine image, and the stereotypical scientist is male, facts which shape perceptions of girls' and women's abilities to be scientists (Easlea 1986). Quite young girls may find it difficult to think of themselves as potential scientists, and in many environments they are actively discouraged by teachers, peers, and parents (Orenstein 1994). The result is that by the end of their secondary education, many girls who might have been interested in scientific careers already lack some of the background education they need to study science in colleges and universities. Stereotypes continue. In an analysis of advertisements in the journal *Science*, Mary Barbercheck argues that "there is more cultural content between the covers of *Science* than we would care to acknowledge" (Barbercheck 2001: 130). Men are more likely to be depicted as scientific heroes than are women; they are also more likely to be depicted as nerds or nonconformists, and more likely to be depicted at play. Words like "simple" and "easy" are more likely to be associated with pictures of women than men, and words like "fast," "reliable," and "accurate" are more likely to be associated with pictures of men than women. And female figures can serve as representations of nature in a way that male figures cannot.

As graduate students, women may be excluded from key aspects of scientific socialization and training. Etzkowitz, Kemelgor and Uzzi (2000) refer to the "unofficial PhD program," that runs alongside the official courses, exams, and research. The unofficial PhD can involve everything from study groups, pick-up basketball games, informal mentoring, and simply ongoing conversations and consultation. Even well-integrated students may be excluded from parts of the unofficial program. One female doctoral student in their study reports: "We would all go to parties together and go and have beer on Friday, but if somebody came in to ask what drying agent to use to clean up THF, they would never ask me. It just wasn't something that would cross their minds" (Etzkowitz et al. 2000: 74). Sexism needs to be studied as part of the everyday texture of research life.

Staying in brings its own set of problems. Being hired into good positions may be more difficult for women, simply because they tend not to look like younger versions of the people hiring them, and possibly because of direct discrimination. A study of the peer-review scores for postdoctoral fellowships by the Swedish Medical Research Council showed that reviewers rated men's accomplishments as overwhelmingly more valuable than apparently equivalent accomplishments by women. Christine Wenneras and Agnes Wold showed that "for a female scientist to be awarded the same competence score as a male colleague, she needed to exceed his scientific productivity by . . . approximately three extra papers in *Nature* or *Science* . . . or 20 extra papers in . . . an excellent specialist journal such as *Atherosclerosis*, *Gut*, *Infection* and *Immunity*, *Neuroscience*, or *Radiology*" (Wenneras and Wold 2001). Thus discrimination can take very blatant forms.

Shirley Tilghman, a very successful molecular biologist and president of Princeton University, argues that the professional pressures and the tenure clock are particularly difficult for women (Tilghman 1998). Tenure, an all-

or-nothing hoop through which young university faculty must jump, typically means during the first six or so years of a university appointment there is no time for anything except teaching and research. That pre-tenure period normally follows several years as a post-doctoral researcher, during which research productivity is essential to gaining a faculty position. Altogether, this professional pressure usually comes at the time when women are most interested in having children and spending time with their young families. Tilghman sees this as an argument against the tenure system: if there were less pressure on those initial years, women might be able to shape their lives differently, having children earlier and putting off some of the advancement of their careers until later.

Since the stratification of science is shaped by cumulative advantage and disadvantage, early careers are particularly important. The problems faced by women as graduate students and young researchers are particularly important for their long-term success. Recent studies at the Massachusetts Institute of Technology and at the University of Michigan (Fox 1991) have shown that young women scientists are often discriminated against, at least in small ways. They tend to have smaller laboratories than male peers hired at the same time, and they tend to have negotiated worse arrangements for start-up grants, for the portion of their research grants that the university takes to cover overhead, for salary, and for teaching duties.

Women are also excluded from informal networks after they have successfully launched careers. Within their departments they tend not to have informal mentors who can help steer them through difficult processes. Outside their departments they tend not to do as much collaborative research as do men. And they tend not to be as well tied in to informal networks that inform them of unpublished research, leaving them more dependent upon the published literature (Fox 1991). While women have moved into science, they are not *of* the community of science (Cole 1981).

While in individual cases such differences can appear trivial, if competition is strong enough they may have a substantial effect. That is, if the cumulative advantages and disadvantages are strong enough, the small irritants that women face, and may even shrug off, can be enough to keep them from getting on (Cole and Singer 1991).

Finally, there is the question of whether women and men learn and do science differently. On the one hand, there have been a number of studies and claims that there is a "math gene" or some similar biological propensity that accounts for boys' better abilities in mathematics and science (e.g. Benbow and Stanley 1980). This claim cannot make sense as other than an ideological statement (e.g. Lewontin 1991), but in addition it runs against recent empirical trends in which boys' and girls' test scores in mathematics are becoming closer and closer (Etzkowitz et al. 2000). On the other hand, a number of feminists and non-feminists have claimed that women and men tend to have distinctive patterns of learning and thinking (e.g. Gilligan 1982; Keller 1985). Women may tend to think in more concrete terms than do men, and are thereby better served by problem-centered learning. They may think in more relational terms than do men, and therefore produce more complex and holistic pictures of their phenomena than do men. This set of issues will be left to chapter 13.

Conclusion

The cumulative advantage and disadvantage hypothesis provides a way of understanding how a nebulous culture can have very concrete effects. For example, science and engineering are and have been dominated by men, and men are still stereotypical scientists and engineers in most fields. Women's participation in science and engineering therefore violates gendered expectations. Women thus find themselves fighting a number of uphill battles. While no one of these may decisively keep women out of the pipeline, eject women from it, or block their progress through it, their compounded effects may be enough to prevent women from moving smoothly into and through it. The cumulative advantage and disadvantage hypothesis provides a way of understanding how women's stories of succeeding and not succeeding in science can be very different, but nonetheless fit a consistent pattern.

With discrimination in focus, it is difficult to see science as efficiently using its resources. Instead of an efficient meritocracy, science looks more like an inefficient old-boy network, in which people and ideas are deemed important and become important at least partly because of their place in various social networks. In the broadest sense, then, elite groups are socially constructed, rather than being mere reflections of natural and objective hierarchies. Undoubtedly, elites produce better knowledge than do non-elites: they have cumulative advantages, and are more likely to be positively reinforced for their work. However, the fact of discrimination raises the question of whether some of the knowledge for which elites are recognized becomes important because of its association with them, and not vice versa.

CHAPTER 5

The Strong Programme and the Sociology of Knowledge

The Strong Programme

In the 1970s a group of philosophers, sociologists, and historians based in Edinburgh set out to understand the *content* of scientific knowledge in sociological terms, developing the "strong programme in the sociology of knowledge" (Bloor 1991 [1976]; Barnes and Bloor 1982; MacKenzie 1981; Shapin 1975). The most concise and best-known statement of the programme is David Bloor's "four tenets" for the sociology of scientific knowledge:

1 It would be causal, that is, concerned with the conditions which bring about belief or states of knowledge. . . .
2 It would be impartial with respect to truth and falsity, rationality or irrationality, success or failure. Both sides of these dichotomies will require explanation.
3 It would be symmetrical in its style of explanation. The same types of cause would explain, say, true and false beliefs.
4 It would be reflexive. In principle its patterns of explanation would have to be applicable to sociology itself.   (Bloor 1991 [1976]: 5)

These much-discussed four tenets represent a bold but carefully crafted statement of a naturalistic, and thus perhaps scientific, attitude toward science and scientific knowledge, extended to technological knowledge as well. Beliefs are akin to objects, and come about for reasons or causes. It is the job of the sociologist of knowledge to understand these reasons or causes. Seen as objects, there is no *a priori* distinction between beliefs that we judge true and those false, or those rational and irrational; in fact, rationality and irrationality are themselves objects of study. And there is no reason to exempt sociology of knowledge itself from sociological study.

S&TS since the strong programme has been centrally concerned with showing how much of science and technology can be accounted for by the *work* done by scientists, engineers, and others. To do so S&TS has emphasized Bloor's symmetry and impartiality strictures, that beliefs judged true and false or rational and irrational should be explained using the same types of resources. This methodological symmetry is a reaction against an unsymmetrical pattern or style of explanation, in which true beliefs require internal, rationalist explanations, whereas false beliefs require external or social explanations. Because it represents a teleological view, this is a variant on the theme of Whig history (chapter 2). Whig history of science rests on the tacit assumption that there is a relatively unproblematic rational route from the material world to correct beliefs about it. That is, Whig history of science rests on a version of foundationalism on which accepted facts and theories ultimately rest on a solid foundation in nature (box 2.2). But as the problems of induction described so far show, there is no guaranteed path from the material world to scientific truths, and no method identifies truths with certainty, so the assumption is highly problematic. Truth and rationality should not be privileged in explanations of particular pieces of scientific knowledge; that is, the same types of factors are at play in the production of truth as in the production of falsity. Since ideology, idiosyncrasy, political pressure, etc., are routinely invoked to explain beliefs thought false, they should also be invoked to explain beliefs thought true.

Although there are a number of possible interpretations of symmetry, in practice symmetry is often equivalent to agnosticism about scientific truths. Bloor's methodological rule tells us to assume that debates are open when we attempt to explain closure. The advantage of such agnosticism comes from its push for ever more complete explanations. The less the analyst takes for granted, the more likely she will have to cast a wide net to give a satisfactory account; too much rationalism tends to make the analyst stop too soon. Establishing the causes of the closure of a debate requires looking at a large number of types of factors, and thus multiple analytic frameworks are valuable for the empirical study of science and technology.

In addition to convincing statements of the potential for sociology of knowledge, strong programmers Bloor (1991) and Barry Barnes (1982; also Barnes, Bloor, and Henry 1996) have offered a useful restatement of the problem of induction. The concept of *finitism* is the idea that each application of a term, classification, or rule requires judgments of similarity and difference. No case is or is not the same as those that came before it, in the absence of a human decision about sameness – though people observe similarities and differences, and make their decisions on the basis of those observations. Terms, classifications, and rules are extended to new cases, but do not simply apply to new cases before their extension.

Actors and observers normally do not feel or see the open-endedness that finitism creates. According to Barnes and Bloor, this is because different kinds of social connections fill most of the gaps between past practices and their extension to new cases. That is, since there is no asocial logic that forces a term, classification, or rule to apply to a new case, what allows for easy application is social forces that push interpretations in one direction or another. This sociological finitism (see box 5.1) opens up a large space for the sociology of knowledge! As Bloor says:

> Can the sociology of knowledge investigate and explain the very content and nature of scientific knowledge? Many sociologists believe that it cannot. . . . They voluntarily limit the scope of their own enquiries. I shall argue that this is a betrayal of their disciplinary standpoint. All knowledge, whether it be in the empirical sciences or even in mathematics, should be treated, through and through, as material for investigation.   (Bloor 1991 [1976]: 1)

Box 5.1   Sociological finitism

The argument for finitism is a restatement of Wittgenstein's argument about rules (box 3.2). Rules are extended to new cases, where extension is a process. Rules therefore change meaning as they are applied. Words and classifications are just special cases of rules.

The application of finitism can be illustrated through an elegant case in the history of mathematics analyzed by Imre Lakatos (1976), and re-analyzed by Bloor (1978). Details are omitted here, but the mathematics is straightforward and helpfully presented, especially by Lakatos. The case concerns a conjecture due to the mathematician Leonard Euler: for polyhedra, $V-E+F=2$, where F is the number of faces, E the number of edges, and V the number of vertices. Euler's conjecture was elegantly and simply proven in the early nineteenth century, and according to the normal image of mathematics that should have been the end of the story. However, quite the opposite occurred. The proof seemed to prompt counterexamples, cases of polyhedra for which the original theorem did not apply!

Problems with the proof and the original theorem led to a variety of different responses. Some mathematicians took the counterexamples as an indictment of the original conjecture; the task of mathematics was then to find a more complicated relationship between V, E, and F that preserved the original insight, but was true for all polyhedra. Other mathematicians took the counterexamples as showing an unacceptable looseness of the category *polyhedra*; the task of mathematics was then to find a definition of *polyhedra* that made Euler's conjecture and its proof correct, and that ruled the strange counterexamples as "monsters." Still others saw the problems as opportunities for interesting classificatory work that preserved the original conjecture and proof while recognizing the interest in the counterexamples.

It seems reasonable to claim that at the time there was no correct answer to the question of whether the counterexamples were really polyhedra. Despite the fact that mathematicians had been working with polyhedra for millennia, any of the responses could have become correct, because the meaning of polyhedra had to change in response to the proof and counterexamples. Sociological finitists claim that what determined how different people extended the term polyhedra was social factors. On Bloor's analysis, the types of societies and institutions in which mathematicians worked shaped their responses, determining whether they saw the strange counterexamples as welcome new mathematical objects with just as much status as the old ones, as mathematical pollution, or simply as new mathematical objects to be integrated into complex hierarchies and orders.

Interest Explanations

The four tenets of the strong programme do not set limits on the resources available for explaining scientific and technological knowledge, and do not establish any preferred styles of explanation. In particular, they do not distinguish between externalist and internalist explanations. Externalist explanations focus on social forces and ideologies that extend beyond scientific and technical communities, whereas internalist explanations focus on forces that are endemic to those communities. The distinction is not perfectly sharp or invariant, nor are many empirical studies confined to one or another side of the divide.

Though statements like Bloor's make it clear that the strong programme is intended to cover both externalist and internalist studies, it was early on strongly associated with the former. This may be because when the strong programme was articulated in the 1970s, historians of science were having historiographical debates about internal and external histories, particularly in the context of Marxist social theory. The strong programme may have slid too neatly into an existing discussion.

Externalist historians or sociologists of science attempt to correlate and connect broad societal events and more narrow intellectual ones. Some difficulties in this task can be seen in an exchange between Steven Shapin (Shapin 1975) and Geoffrey Cantor (Cantor 1975). Shapin's article, itself a response to a piece by Cantor, argues that the growth of interest in phrenology in Edinburgh in the 1820s was related to a heightened class struggle there. The Edinburgh Phrenological Society and its audiences for lectures on phrenology were dominated by members of the lower and middle classes. Meanwhile, the Royal Society of Edinburgh, which had as members many of the strongest critics of phrenology, was dominated by the upper classes.

Reasons for these correlations are complex and somewhat opaque, but Shapin claims that they are related to connections between phrenology and reform movements.

Cantor makes a number of criticisms of Shapin's study: (1) Class membership is not a clear-cut matter, and so the membership in the societies in question may not be easily identified as being along class lines. In addition, on some interpretations there was considerable *overlap* in membership of the two societies, thus begging the question of the extent to which the Phrenological Society could be considered an outsider's organization. (2) Shapin does not define "conflict" so as to demonstrate that there was significantly more conflict between classes in the Edinburgh of the 1820s than there had been at some other time. (3) While the overall picture of membership of the two societies may look different, they had very similar percentages of members coming from some professional groups. To make this point vivid, Cantor calls for a social explanation of the *similarity* of composition of the two societies. A correlation may not be evidence for anything.

While Cantor's criticisms of Shapin are specific to this particular account, they can be applied to other interest-based accounts. Many histories of the 1970s and 1980s follow the same pattern. They identify a scientific controversy in which the debaters on each side can be identified. They identify a social conflict, the sides of which can be correlated to the sides of the debate. And finally, they offer an explanation to connect the themes of the scientific debate and those of the social conflict (e.g. MacKenzie 1978; Jacob 1976; Rudwick 1974; Farley and Geison 1974; Shapin 1981; Harwood 1976, 1977).

Exactly the same problems may face many internalist accounts, but are not so apparent because the social divisions and conflicts often seem proper to science and technology, and as such are more immediately convincing as causes of beliefs. Conflicts between physicists with different investments in mathematical skills (Pickering 1984), between natural philosophers with different models of scientific demonstration (Shapin and Schaffer 1985), or between proponents of different methods of making steel (Misa 1992), involve much more immediate links between interests and beliefs, because the interests are apparently internal to science and technology.

Steve Woolgar (1981) has developed a further criticism of interest-based explanations. Analysts invoke interests to explain actions even when they cannot display a clear causal path from interests to actions. To be persuasive, then, analysts have to isolate a particular set of interests as dominant, and independent of the story being told. However, there are indefinitely many potential interests capable of explaining an action, so any choice is underdetermined. Woolgar's larger project is a critique of "social realism," the position that aspects of the social world are fully determinate even if aspects of the natural world are not. Woolgar, pushing forward the reflexive part of the strong programme, points out that accounts in science and technology studies (S&TS) rhetorically construct aspects of the social world, in this case interests, in exactly analogous ways as scientists construct aspects of the natural world. S&TS needs to figure out how to make social reality and natural reality symmetrical, or to justify their lack of symmetry. Most prominent in attempting the former is actor-network theory (Latour 1987; Callon 1986). Most prominent in attempting the latter is the program of methodological relativism (Collins and Yearley 1992).

Interest-based patterns of analysis thus face a number of problems: (1) analysts tend to view the participants in the controversy as two-dimensional characters, having only one type of social interest, and a fairly simple line of scientific thought; (2) they tend to make use of a simplified social theory, isolating relatively few conflicts and often simplifying those that are employed; (3) it is difficult to show that there is a real link between social group membership and belief; and (4) interests are usually taken as fixed, and society as stable, even though these are as constructed and flexible as are the scientific results to be explained.

Despite these problems, S&TS has not abandoned interest-based explanations – they are too valuable to be simply left aside. Interest explanations are essentially rational choice explanations, in which actors do things in order to meet their goals. Rational choices need to be situated in a context in which certain goals are highlighted, and the choices available to reach those goals are narrowed. The difficult theoretical problems are thus answered in practical terms, by more detailed and cautious empirical work. Researchers in S&TS have paid increasing attention to scientific and technical cultures, especially material cultures, and how those cultures shape options and choices. They have emphasized clear internal interests, such as interests in particular approaches or theories. As a result, researchers in S&TS have shown how social, cultural, and intellectual matters are not distinct; instead, intellectual issues have social and cultural ones woven into their very fiber. Finally, while situating these choices involves rhetorical work on the part of the analyst, this is just the sort of rhetorical work that any explanation requires. Woolgar's critique is not so much of interests, but is a commentary on explanation more broadly (Ylikoski 2001).

Box 5.2   Testing technologies

A test is only as revealing as it is representative. A test of a technology shows its capabilities only to the extent that the circumstances of the test are the same as "real-world" circumstances. According to the finitist argument, though, this issue is always open to interpretation (Pinch 1993a). To make this point concrete we might ask with Donald MacKenzie (1989), how accurate are ballistic missiles? This is an issue of enormous importance, not least to the militaries and governments that control the missiles. As a result, there have been numerous tests of unarmed ballistic missiles. Although the results of these tests are mostly classified, some of the debates around them are not.

As MacKenzie documents, the critics of ballistic missile tests point to a number of differences between test circumstances and the presumed real circumstances in a nuclear war. For example, in the United States, at least until the 1980s when MacKenzie was doing his research, most intercontinental ballistic missiles had been launched from Vandenberg Air Force Base in California to the Kwajalein Atoll in the Marshall Islands; thus the variety of real launch sites, targets, and even weather was missing from the tests. In a test, an active missile is selected at random from across the US, sent to Texas where its nuclear warhead is removed and replaced by monitoring equipment, and where the missile is wired to be blown up in case of malfunction, and finally sent on to California. To make up for problems introduced in transportation, there is special maintenance of the guidance system prior to the launch. It is then launched from a silo that has itself been well maintained and well studied. Clearly there is much for critics of the tests to latch on to. A retired general says that "About the only thing that's the same [between the tested missile and the Minuteman in the silo] is the tail number; and that's been polished" (MacKenzie 1989: 424). Are the tests representative, then? Unsurprisingly, interpretations vary, and they are roughly predictable from the interests and positions of the interpreters.

Trevor Pinch (1993a) gives a general and theoretical treatment of MacKenzie's insights on technological testing. He points out that whether tests are prospective (testing of new technologies), current (evaluating the capabilities of technologies in use), or retrospective (evaluating the capabilities of technologies typically after a major failure), in all three of these situations projections have to be made from test circumstances on to real circumstances. Judgments of similarity have to be made, and are potentially defeasible. Therefore, it is only via human judgment that we can project what technologies are capable of, whether in the future, in the present, or in the past.

Knowledge, Practices, Cultures

From the perspective of many of its critics, the strong programme rejects truth, rationality, and the reality of the material world. In part such interpretations come from the association of the strong programme with too-simple externalist histories. As Barry Barnes, David Bloor, and John Henry (1996) argue, it is difficult to make sense of such perspectives and readings. The strong programme does not reject any of these touchstones, but rather shows how by themselves truth, rationality, and the material world have limited value in understanding why one scientific claim is believed over another.

For other critics, the strong programme retains too much commitment to truth and the material world. Barnes, Bloor, and Henry (1996) reject what they see as idealist tendencies in S&TS. Some people, notably Harry Collins, see their arguments on this issue as weakening the commitment to methodological agnosticism about scientific truths. Whether this is right remains an open question.

Finally, the strong programme has been criticized for being too committed to the reality and hardness of the social world: it is seen as adopting a foundationalism in the social world to replace the foundationalism in the material world that it rejects. As Woolgar argues, there is no reason to see interests and their balance as less malleable than the positions themselves. The critique of interests has been amplified by Bruno Latour (1987; also Pickering 1995), who argues that interests are translated and modified in the production of scientific knowledge and technological artifacts, and thus that society, science, and technology are produced together, and by the same processes – this results in a "supersymmetry" (Callon and Latour 1992).

Since the 1970s, the pendulum in S&TS, and elsewhere in the humanities and social sciences, has swung from emphasizing structures to emphasizing agency (e.g. Knorr Cetina and Cicourel 1981). Although its statements leave this issue entirely open, the strong programme was often associated with structuralist positions. Thus as a philosophical underpinning for S&TS the strong programme has been supplemented by others: constructivism (e.g. Knorr Cetina 1981), the empirical program of relativism (Collins 1991 [1985]), actor-network theory (e.g. Latour 1987), symbolic interactionism (e.g. Clarke 1990), and ethnomethodology (e.g. Lynch 1991). For the most part, we will avoid anatomizing theoretical positions and disputes in this book, except as they directly inform its main topics, but readers interested in philosophical and methodological underpinnings might look at these works and many others to see some of the contentious debates around foundational issues.

The strong programme provided an argument that one can study the content of science and technology in social and cultural terms, and thus provided an argument for the possibility of S&TS. As the field has developed, scientific and technological practices themselves became interesting, not just as steps to understanding knowledge. Science as an activity has become a productive locus of study. And key epistemic concepts – such as experimentation, explanation, proof, and objectivity – understood in terms of the roles they play in scientific practice, have become particularly interesting in S&TS. Thus the strong programme's singular attention to explaining pieces of knowledge in social terms now seems a partial perspective on the project of understanding science and technology, even if it is a crucial foundation.

CHAPTER 6

The Social Construction of Scientific and Technical Realities

What Does "Social Construction" Mean?

The term *social construction* started to become common in science and technology studies (S&TS) in the late 1970s (e.g. Mendelsohn 1977; Van den Daele 1977; Latour and Woolgar 1979). Since then, important texts have claimed to show the social construction (or simply the construction) of facts, knowledge, theories, phenomena, science, technologies, and societies. Social constructivism, then, has been a convenient label for what holds together a number of different parts of S&TS. And social constructivism has been the target of fierce arguments by historians, philosophers, and sociologists, usually under the banner of *realism.*

For S&TS social constructivism provides three important assumptions, or perhaps reminders. First is the reminder that science and technology are importantly *social.* Second is the reminder that they are *active* – the construction metaphor suggests activity. And third is the reminder that science and technology do not provide a direct route from nature to ideas about nature, that the products of science and technology are *not themselves natural.* While these reminders have considerable force, they do not come with a single interpretation. As a result, there are many different "social constructions" in S&TS, with different implications. This chapter offers some categories for thinking about these positions, and the realist positions that stand in opposition to them.

The classification here is certainly not the only possible one. For example, one of Ian Hacking's goals in a recent book-length discussion of constructivism (Hacking 1999) is to chart out the features of a strong form of social constructivism, to provide the sharpest contrast to realism. The features he identifies are: the *contingency* of facts, *nominalism* (discussed below) about kinds, and *external explanations* for stability. On his analysis, a social constructivist account of, say, an established scientific theory, will tend toward the position that: the theory was not the only one that could have become established, the categories used in the theory are human impositions rather than in nature, and the reasons for the success of the theory are not evidential reasons.

The goals of this chapter are somewhat different. Rather than providing a single analysis in terms of a sharp contrast to realism, the chapter pulls social constructivism apart into a number of different types of claims. Some of these are clearly anti-realist, and some are not. While there are some general affinities between these claims, they are also interesting in their specificity.

Although most of the forms of constructivism described below are opposed to forms of realism, there is less need for an analogous list of realisms. This does not mean that realism is any more straightforward. Realism typically amounts to an intuition that truths are more dependent upon the world than upon the people who articulate those truths. There is a way that the world is, and it is possible to discover and represent it. Among realists, though, there are disagreements over exactly what science realistically represents, and over what it means to realistically represent something. There are disagreements about whether the issue is fundamentally one about knowledge or about things. And there is no good account of the nature of reality, what makes real things real – for this reason, realism is probably less often a positive position than a negative one, articulated in opposition to one or another form of anti-realism.

1   The construction of social reality

Ideas of social construction have many origins in classic sociology and philosophy, from analyses by Karl Marx, Max Weber, and Émile Durkheim, among others. S&TS imported the phrase "social construction" from Peter Berger and Thomas Luckmann's *The Social Construction of Reality* (1966), an essay on the sociology of knowledge. That work provides a succinct argument for why the sociology of knowledge studies the social construction of reality:

> Insofar as all human "knowledge" is developed, transmitted and maintained in social situations, the sociology of knowledge must seek to understand the processes by which this is done in such a way that a taken-for-granted "reality" congeals for the man in the street. In other words, we contend that *the sociology of knowledge is concerned with the analysis of the social construction of reality.*   (Berger and Luckmann 1966: 3; italics in original)

The primary reality in which Berger and Luckmann are interested, though, is social reality, the institutions and structures that come to exist because of people's actions and attitudes. These features of the social world exist because significant numbers of people know them to exist, and act accordingly. Rules of polite behavior, for example, have real effects because people act on them and act with respect to them. "Knowledge about society is thus a realization in the double sense of the word, in the sense of apprehending the objectivated social reality, and in the sense of ongoingly producing this reality" (Berger and Luckmann 1966: 62). Features of the social world exist because they are "independent of our own volition," because we cannot "wish them away" (1966: 1). And as long as people act with respect to them, they reinforce these social realities (see also Searle 1995).

The central point here, or at least the central insight, is one about the metaphysics of the social world. To construct X we need only that: (a) knowledge of X encourages behaviors that reduce other people's ability to act as though X does not exist; (b) there is reasonably common knowledge of X; and (c) there is transmission of knowledge of X. Given these conditions, X cannot be "wished away," and so it exists. For example, gender is real, because it is difficult not to take account of it. Gender structures create constraints and resources with which people have to reckon. As a result, treating people as gendered tends to create gendered people. Genders have causal powers, which is probably the best sign of reality that we have. At the same time, they are undoubtedly not simply given by nature, as historical research and divergences between contemporary cultures show us.

For S&TS, knowledge, methods, epistemologies, disciplinary boundaries, and styles of work are all key features of scientists' and engineers' social landscapes. To say that these objects are socially constructed in this sense is simply to say that they are *real* social objects, though contingently real. Ludwik Fleck, a key forerunner of S&TS, already makes these points in his *Genesis and Development of a Scientific Fact* (1979 [1935]). A scientist or engineer who fails to account for a taken-for-granted fact in his or her actions may encounter resistance from colleagues, which shows the social reality of that fact. To point this out is in no way to criticize science and technology. The social reality of knowledge and of the practices around knowledge is a precondition of progress. If nothing is reasonably solid, then there is nothing on which to build.

S&TS has tended to add an active dimension to this metaphor, looking at processes of social construction. Claims do not just spring from the subject matter into acceptance, via passive scientists, reviewers, and editors. Rather, it takes work for them to become important. For example, Latour and Woolgar (1979) chronicle the path of the statement "TRF (thyrotropin releasing factor) is Pyro-Glu-His-Pro-NH_2," as it moves from being *near nonsense* to *possible* to *false* to *possibly true* to a *solid fact*. Along the way, they chart out the different operations that can be done on a scientific paper: ignoring it to detract, citing it positively, citing it negatively, questioning it (in stronger and weaker ways), and ignoring it because it is universally accepted. Scientists, and not just science, construct facts.

Box 6.1   The social construction of the discovery of the laws of genetics

Discoveries are important to the social structure of science, because more than anything else, recognition is given to researchers for what they discover. This recognition motivates priority disputes over discoveries, as well as sometimes-fierce struggles for priority (Merton 1973). But it is often difficult to pinpoint the moment of discovery. Was oxygen discovered when Joseph Priestley created a relatively pure sample of it? Or when Antoine Lavoisier followed up on Priestley's work with the account of oxygen as an element? Or at some later point when an account of oxygen was given that more or less agrees with our own (Kuhn 1970a)? Such questions lead Augustine Brannigan (1981) to an *attributional* model of discoveries: discoveries are not events by themselves, but are rather events retrospectively recognized as origins.

Brannigan's discussion of Mendelian genetics nicely illustrates the attributional model. On the standard story, Gregor Mendel was an isolated monk who performed ingenious and simple experiments on peas, to learn that heredity is governed by paired genes that are passed on independently, and can be dominant or recessive. His 1866 paper was published in an obscure journal and was not read by anybody who recognized its importance until 1900, when Hugo de Vries, Carl Correns, and Erich Tschermak came upon it in the course of their own similar studies. Against this story, Brannigan shows that Mendel's paper was reasonably well cited before 1900, but in the context of agricultural hybridization. In that context, its main result, the 3:1 ratio of characteristics in hybrids, was already known. Mendel was read as replicating that result, and offering a formal explanation of it. Indeed, Mendel does not seem to recognize his results or theory as a momentous discovery, judging by his presentation of them.

Though de Vries had read Mendel's paper before 1900, possibly much before, the first of his three publications on his own experiments and the new insights on genetics makes no mention of it. Correns, who was pursuing a similar line of research, read de Vries's first publication, and quickly wrote up his own, labeling the discovery "Mendel's Law." Recognizing that he had lost the race for priority, Correns assigned it to an earlier generation – though Brannigan argues that had Correns read papers from one, two, or three generations earlier still the title could easily have gone to any of a number of potential discoverers. De Vries's second and third publications accept the priority of Mendel in awkward apparent afterthoughts, but grumble about the obscurity of Mendel's paper.

> Mendel appears to have become the discoverer of his laws of ge-
> netics as a result of a priority dispute. Because of that dispute, his
> work on hybridization was pulled out of that context and made to
> speak to the newly important questions of evolution. The discovery of
> Mendel's laws, then, took place in 1866, but only because of the events
> of 1900.

2   The construction of things and phenomena

Not only representations and social realities are constructed. Perhaps the most novel of S&TS's constructivist insights is that many of the things that scientists and engineers study and work with are non-natural. The insight is not new – it can be found in Gaston Bachelard's (1984 [1934]) concept of *phenomenotechnique*, and may even be found in the work of Francis Bacon written in the early seventeenth century – but it is put forcefully by researchers in S&TS writing in the 1980s.

Karin Knorr Cetina writes that "nature is not to be found in the laboratory" (Knorr Cetina 1981: 4). "To the observer from the outside world, the laboratory displays itself as a site of action from which 'nature' is as much as possible excluded rather than included" (Knorr Cetina 1983: 119). For the most part, the materials used in scientific laboratories are already partly prepared for that use, before they are subjected to laboratory manipulations. Substances are purified, and objects are standardized and even enhanced. Chemical laboratories buy pure reagents, geneticists might use established libraries of DNA, and engineered animal models can be invaluable.

Once these objects are in the laboratory, they are manipulated. They are placed in artificial situations, to see how they react. They are subjected to "trials of strength" (Latour 1987) in order to characterize their properties. In the most desirable of situations scientists create phenomena, new stable objects of study that are particularly interesting and valuable (Bogen and Woodward 1988) – Hacking writes that "even those who construct the longest lists will agree that most of the phenomena of modern physics are manufactured" (Hacking 1983: 228).

The result of these various manipulations is that knowledge derived from laboratories is knowledge about things that are distinctly non-natural. These things are constructed, though it is a hands-on and fully material form of construction. We will return to this form of construction in chapter 15.

In terms of technology, there is nothing striking about this observation. Whereas sciences are presumed to display the forms of nature exactly as they are, technology gives new shape and form to old materials, making objects that are useful and beautiful. The fact that technology involves hands-on and material forms of construction, leaving nature behind, is entirely expected.

3   The scientific and technological construction of material and social environments

Scientific facts and technological artifacts can have substantial impacts on the material and social world – that is the source of much of the interest in them. As such, we can say that science and technology contribute to the construction of many environments.

The effects of technology can be enormous, and can be both intended and unintended. The success of gasoline-powered automobiles helped to create suburbs and the suburban lifestyle, and to the extent that manufacturers have tried to increase the suburban market, these are intentional effects. Similarly, the shape of computers, computer programs and networks are created with their social effects very much in mind: facilitating work in dispersed environments, long-distance control, or straightforward military power (Edwards 1997).

Science, too, shapes the world. Research into the causes of gender differences, for example, has the effect of naturalizing those differences. And there is tremendous public interest in this area, so biological research on genes linked to gender, on the gender effects of hormones, and on brain differentiation between men and women tends to be well reported (Nelkin and Lindee 1995). More often than not, it is reported to emphasize the inevitability of stereotypical gendered behavior, and there is good reason to expect that this has substantial effects on gender itself.

Science also shapes policy. Government actions are increasingly accountable to scientific evidence. Almost no action can be undertaken unless some claim can be made that it is supported by a study, whether it is in areas of health, economy, environment, or defense. Scientific studies, then, have at least some effect on public policies, which have at least some effect on the shapes of the material and social world. Science, as well as technology, then, contributes to the construction of our environments.

4   The construction of theories

The most straightforward use of the social construction metaphor in S&TS describes scientists and engineers constructing accounts, models, and theories, on a basis of data, and methods for moving from data to representations. However, given the problems of induction we have seen, this process cannot be a purely methodical one. There is no mechanical way to develop good theoretical pictures on a basis of finite amounts of data, no direct route from nature to accounts of nature. Therefore, representations of nature are connected to nature, but do not necessarily correspond to it in any strong sense. Science is constructive in a geometrical sense, making patterns appear given the fixed points that practice produces.

That science constructs representations on top of data is roughly the central claim made by logical positivism (see chapter 1). For positivism, there is an essential contingency to scientific theories and the like. For any good scientific theory, one can create others that do equally good jobs of accounting for the data. Therefore, we should not believe theoretical accounts, if to believe them means committing ourselves to their truth. Bas van Fraassen, whose work is positivist in spirit, has developed a position he calls "constructive empiricism," in which we can see the work of the metaphor. He says, "I use the adjective 'constructive' to indicate my view that scientific activity is one of construction of rather than discovery: construction of models that must be adequate to the phenomena, and not discovery of truth concerning the unobservable" (van Fraassen 1980: 5).

For positivists, the contingency of scientific representations is largely eliminated by prior decisions about frameworks; what made it *logical* positivism was the assumption that scientists make decisions about the logical frameworks within which they work. For the analogous form of constructivism in S&TS, though, there is no top-down way of eliminating the contingency of scientific representations. Karin Knorr Cetina's early use of the term "construction" may owe something to van Fraassen, but to explain why particular theories solidify she looks to established practices, earlier decisions, the extension of concepts to new applications, tinkering, and local contingencies (Knorr 1977, 1979; Knorr Cetina 1981).

Controversies easily show the value of bottom-up accounts of contingency (see chapter 10). By definition, scientific and technical controversies display alternative representations, alternative attempts to construct theories and the like. They can also display some of the forces that lead to their closure. For example, the choice between Newton's and Leibniz's metaphysics may have been related to their relative political circumstances (Shapin 1981), and the choice between the neural network approach and the formal approach to artificial intelligence was decided by a combination of forceful rhetoric and the demands of military funding (Guice 1998). Scientific and technological theories, then, are constructed on top of data, but are not implied by those data.

Box 6.2   Realism and empiricism

The question of whether one should believe scientific theories, or merely see them as good working tools, has been one of the most prominent questions in the philosophy of science since the early twentieth century. Most philosophers agree that science's best theories are impressive in their accuracy, but there is real disagreement about whether that provides grounds for believing in the truth of those theories, when the theories invoke unobservable entities or processes.

The classic *empiricist* argument against truth starts from the claim that all of the evidence for a scientific theory is from empirical data. Therefore, given two theories that make exactly the same predictions, there can be no empirical evidence to tell the difference between the two. But any theory we hold is only one of an infinite number of empirically equivalent theories, so there is no reason to think that it is true. Truth, then, is a superfluous concept (e.g. van Fraassen 1980; Misak 1995).

Scientific realists have a number of responses. They can challenge the starting assumption, and argue that empirical data are not the only evidence one can have for a theory. As a result, they can challenge the assumption that there are typically an infinite number of equivalent theories, because only a few theories are truly plausible. And they can argue that there is no way to make sense of the successes of science without reference to the at-least approximate truth of the best theories (see, e.g., Leplin 1984; Papineau 1996). One of the best-developed versions of the latter argument is due to Richard Boyd (e.g. 1985; see Sismondo 1996 for a summary). Boyd argues that we have good reason to believe "what is implicated in instrumentally reliable methodology" (Boyd 1990: 186). The truth of background theories is the best explanation of the success of scientific methods. Boyd's strategy is to focus not on how successful theories are at making predictions or accounting for data, but on how successful background theories are in shaping research, which produces reliable theories.

5   Heterogeneous construction

Successful technological work draws on multiple types of resources, and simultaneously addresses multiple domains, a point that will be developed in chapter 7. The entrepreneurial engineer faces technical, social, and economic problems together, and has to bind solutions to these problems together in a configuration that works. In helping to construct an artifact, then, the engineer is simultaneously helping to construct knowledge, social realities, things, and the material and social world, in the senses described above. While these senses can be parceled out analytically, in practice they are bound together.

Heterogeneous engineering is the term John Law (1987) gives to the work that builders of technologies need to do. They have to simultaneously build their artifacts, and build environments in which they can function – and typically, neither of these activities can be done on their own. Technologists need to combine raw materials, skills, knowledge, and capital, and to do this they must enroll any number of actors, not all of whom may be immediately compatible. Thus technologists have the task of building stable networks involving diverse components.

Scientific work is analogously heterogeneous. Actor-network theory (see chapter 7) is a theory of "technoscience," in which scientists and engineers are separated only by traditional disciplinary boundaries. Like engineers, scientists construct networks, the larger and more stable the better – stable networks create an orderly world, and so the construction of networks is the construction of order. Science's networks are heterogeneous in the sense that they combine isolated parts of the material world, laboratory equipment, established knowledge, patrons, money, institutions, etc. These actors together create the successes of technoscience, and no one piece of a network can wholly determine the shape of the whole.

What we might call *heterogeneous construction* is the simultaneous shaping of the material and social world, to make them fit each other – it is a process of "co-construction" (Taylor 1995). Heterogeneous construction can involve all of the other types of construction mentioned to this point, combining the construction of accounts with the construction of parts of social reality, with the construction of phenomena and the construction of the broader environment. In a review of policy around climate change, Simon Shackley and Brian Wynne (1995) invoke the notion of "mutual construction" to describe how the criteria for good climate science are shaped by policy concerns, and the criteria for good policy are being shaped by climate science. Or, in a different way, Geoffrey Bowker and Susan Leigh Star (2000) describe how ideas and practices concerning health and disease are bent around standardized classifications of diseases and medical interventions. Classification systems are not neutral on the things that they classify.

Box 6.3   The development of the Pap smear

Monica Casper and Adele Clarke (1998), working in the "social worlds" tradition, show how the Pap smear became the "right tool for the job" of screening for cervical cancer through a process that we could see as heterogeneous engineering. The Pap smear has its origins in George Papanicolaou's study, published in 1917, on vaginal smears as indicators of stages in the estrous cycle of guinea pigs. Over the following decade, Papanicolaou investigated other uses of the smears, eventually discovering that he could detect human cancer cells. His original presentation of that finding, in 1928, met with little enthusiasm: the results were not convincing, pathologists were not used to looking at free-floating cells, and gynecologists were uninterested in cancer. Papanicolaou abandoned the smear as a cancer test for another decade. Leaving the history of the test aside, there are problems with the test itself. The Pap smear faces "chronic ambiguities" in the nature of cancer, in the classification of normal and abnormal cells, and in the reading of slides. As a result it has a false negative rate (it fails to detect cancerous and precancerous cells) of between 15 percent and 50 percent of cases, depending upon the circumstances. Given the problems of its initial reception, and the problems with established versions of the test, how did it become such a successful test, in the process contributing to saving the lives of thousands, and perhaps millions, of women?

According to Casper and Clarke, the smear's initial failure was overcome with the arrival of new powerful actors, such as the American Cancer Society, which created an environment that could support the tinkering necessary to address the test's problems. Its later problems were solved by various social and material adjustments. The test became less expensive by gendering the division of labor. Specially trained technicians, predominantly female, could be paid considerably less than the predominantly male pathologists. In some cases the technicians could be paid on a piecework basis, and could do some of their work from home. These innovations could allow the suitable volumes of the test to be performed for it to become an effective screening test. That volume, however, meant that technicians suffered from eyestrain, and from the combination of low status and high levels of responsibility on matters of life and death. There have been other cost-reduction efforts. The automation of record-keeping helped to reduce the cost of Pap smears, and also to make them more useful as screening tests for large populations of women (see Singleton and Michael 1993). High rates of incorrect readings of the tests have created public pressure for rating and regulating the labs performing them. Women's health activists have prompted governments to take seriously the conditions under which Pap smears are read, and the number of smears read by a technician each day, reducing the number of "Pap mills." High rates of incorrect readings have also prompted the negotiation of local orders. Rather than strictly following standard classification schemes, labs sometimes work closely with physicians and clinics, receiving information about the health of the women whose smears they read. Labs thus develop local techniques and classifications which, in clinical tests, appear to have better success rates.

6 The construction of kinds

It has been a long-standing philosophical question whether natural kinds are part of the non-human world or are only part of human classification. (The debate is usually put in terms of "universals," the abstractions that range over individual objects, like redness.) *Nominalists* believe that kinds are human impositions, even if people find it relatively easy to classify objects similarly. *Realists* believe that kinds are real features of the world, even if their edges may be fuzzy and their application somewhat conventional. For nominalists, the only real things are individual objects. Given how difficult it is to make sense of the reality of general properties – Where do they exist? How do they apply to individual objects? – nominalists prefer to make them entirely mental and linguistic phenomena. For realists, the world has to contain more than mere objects. Given how difficult it is to make sense of a world without real general divisions – Do we not discover features of the world? If they were not real, why would kinds have any value? – realists accept that kinds are external to people.

In S&TS, nominalism is one way of cashing out the construction metaphor. If kinds are not features of the world, then they are constructed. To the extent that they are constructed by groups of people, they are socially constructed. Since science is the most powerful of institutions that classify things, science is engaged in the social construction of reality. We can see statements of this in works by Thomas Kuhn (e.g. 1977) and proponents of the strong programme (e.g. Barnes, Bloor, and Henry 1996): sociological finitism (see chapter 5) is at least a weak version of nominalism.

7 The construction of nature

It is only a short step from nominalism to the strongest form of the social construction metaphor in S&TS, the claim that representations directly shape their objects. According to this form of constructivism, when scientists create agreement on a claim, they literally make the claim true. The world follows agreement, not the other way around. Similarly, when engineers create agreement on what the most efficient solution to a problem is, they literally make that solution the most efficient one. Mind, in this case a social version of mind, is prior to nature; the way it classifies and otherwise describes the world becomes literally true. The position bears some similarity to Kant's idea that humans impose some structures on the world, so we can call it neo-Kantian constructivism.

John Guillory (2002) sees this constructivism as "the spontaneous philosophy of the critics," arguing that there are structural causes of its popularity, based in, among other things, the lack of institutional power of the social sciences and humanities. At the level of reasons rather than causes, neo-Kantian constructivism gains its plausibility from two facts. First is the fact that the natures of things are not directly available to us without representations, that we do not have independent access to the way the world is. When scientists, or anybody else, come to agreement about the constitution of something, they do so in response to sense experiences, more mediated information, and arguments. Since even sense experiences are themselves responses to things, there is no direct access to the natures of things.

Second, disagreement is the rule, not the exception. Detailed studies of science and technology suggest that there is a large amount of contingency in our knowledge before it stabilizes. Yet natural and technological objects are relatively well behaved once scientists and engineers come to some agreement about them. On a closely related note, Albert Einstein says, "Science as something already in existence, already completed, is the most objective, impersonal thing that we humans know. Science as something coming into being, as a goal, is just as subjectively psychologically conditioned as are all other human endeavors" (quoted in Kevles 1998: 15). We need to account for this change.

Because of the initial contingency, we cannot say that there is a way the world is that guarantees how it will be represented. And therefore, we might question the priority of objects over their representations. Steve Woolgar probably comes closest to the neo-Kantian position in S&TS, when he argues that contingency undermines realist assumptions:

> The existence and character of a discovered object is a different animal according to the constituency of different social networks. . . . Crucially, this variation undermines the standard presumption about the existence of the object prior to its discovery. The argument is not just that social networks mediate between the object and observational work done by participants. Rather, the social network constitutes the object (or lack of it). The implication for our main argument is the inversion of the presumed relationship between representation and object; the representation gives rise to the object. (Woolgar 1988: 65)

Neo-Kantian constructivism is difficult to accept. The modest version of its central claim that was put forward by Immanuel Kant was attached to an individualist epistemology: individuals impose structure on the world as they apprehend it. For the Kantian, the individual's isolation from the material world suggests that it makes no sense to talk of anything lying beyond phenomena, which are in part constituted by people's frameworks and preconceptions. However, S&TS's neo-Kantianism should not be so modest, because S&TS emphasizes the *social* character of scientific knowledge. If what is at issue are the representations made by groups of people, the neo-Kantian claim appears less motivated. How does consensus affect material reality? Or how do the convictions of authorities carry weight with the world that the convictions of non-authorities do not? How does consensus *cause* changes in the material world?

If neo-Kantian constructivism is true, for example, then representation acts at a distance in a way that is normally thought impossible. Successful representation makes changes, and may even create what it represents, even though there are no causal connections from representation to represented. Neo-Kantian constructivism therefore violates some fundamental assumptions about cause and effect. It is for this reason that authors such as Latour argue that social constructivism is implausible for more than a second (e.g. Latour 1990).

There are also political concerns about neo-Kantian constructivism. Feminists point out (see chapter 13) that science's images of women are sometimes sexist, particularly in that they are quick to naturalize gender. If neo-Kantian constructivism were right, then, while feminist critics could attempt to change science's constructions of women, they could not reject them as false – scientific consensus is by definition true. Similarly, environmentalists have a stake in the reality of nature aside from constructions of it. While they can attempt to change dominant views on the resilience of nature, they could not reject them as false (Grundmann and Stehr 2000; Takacs 1996).

In light of the obvious truth of at least some of the other versions of constructivism described above, though, neo-Kantian constructivism may represent a decent approximation, and may be methodologically valuable. So while claims about the "social construction of reality" can sometimes look suspect, they may amount to little more than metaphor or sweeping language. Even the political problems with neo-Kantianism may be unimportant, if the language is understood correctly (Burningham and Cooper 1999). Claims about the "social construction of reality" may draw attention to contingency in science and technology, and therefore lead researchers to ask about the causes of contingency. Seen as a metaphor, this strong neo-Kantianism can be a valuable tool in the investigation of science and technology.

Richness in Diversity

At the same time that the term has become common in S&TS, social construction talk has taken off in the humanities and social sciences in general, so much so that the philosopher Ian Hacking asks in the title of a book, *The Social Construction of What?* (Hacking 1999). Genders, emotions, identities, and political movements are only a few of the things to which social construction talk has been applied.

S&TS is partly responsible for the explosion of talk of social construction. Because scientific knowledge is usually seen as simply reflecting the natural world, and scientists must therefore be relatively passive in the creation of that knowledge, the claim that scientific realities are socially constructed is a very strong one. As a result, S&TS's constructivist claims have been influential. This can be seen in the explicit use of constructivist resources from S&TS in such fields as psychology, geography, environmental studies, education, management, cultural studies, and even accounting.

However, the diversity of claims about the social construction of reality means that constructivism in S&TS cannot be any neat theoretical picture. Instead, constructivism provides reminders of the points with which we began this chapter, that science and technology are social, that they are active, and that they do not take nature as it comes.

Some of these forms of constructivism are controversial on any reading, and all of them are potentially controversial in the details of their application. But given their diversity it is also clear that even the staunchest of realists cannot dismiss constructivist claims out of hand. Constructivism is the study of how scientists and technologists make socially situated knowledges and things. Such studies can even show how scientists build good representations of the material world, in a perfectly ordinary sense. As recognized by some of the above strains of constructivism, science gains power from, among other things, its ability to manipulate nature and measure nature's reactions, and its ability to translate those measurements across time and space to other laboratories and other contexts. Laboratory and other technologies thus contribute to objectivity and objective knowledge. As a result, constructivist S&TS may even *support* a version of realism, though not the idea that there is unmediated knowledge of reality, nor the idea that there is one complete set of truths.

CHAPTER 7

Actor-Network Theory

Actor-Network Theory: Relational Materialism

Actor-network theory (ANT) is the name given to a framework originally developed mostly by Michel Callon (e.g. 1986), Bruno Latour (e.g. 1987), and John Law (e.g. 1987). ANT has its origins in an attempt to understand science and technology – or *technoscience* in Latour's (1987) terminology, since on this account science and technology involve the same processes. It is, though, a general social theory centered on technoscience, rather than just a theory of technoscience.

The theory represents the work of technoscience as the creation of larger and stronger networks. Part of this is in straightforward analogy to traditional analyses of power politics: just as the political actor strives to put together alliances that allow him or her to maintain power, so do scientists and engineers. However, the actors of ANT are heterogeneous in that they include both human and non-human actors, with no important distinction between them. Both humans and non-humans have *interests* that need to be accommodated, and that can be managed and used. Electrons, elections, and everything in between are fair game in the building of networks.

Michel Callon (1987), for example, describes the effort of a group of engineers in Electricité de France (EDF) to introduce an electric car in France. EDF's engineers acted as "engineer-sociologists" in the sense that they simultaneously articulated a vision of fuel cells for these new cars, of French society into which electric cars would later fit, and of much between the two – engineering is never complete if it stops at the obvious boundaries of engineered artifacts. Not only the EDF actors did engineering-sociology; their opponents at Renault, who were committed to internal combustion engines, criticized both the technical details and the social feasibility of EDF's plans. The engineering and the sociology are inseparable, because neither the technical vision nor the social vision will come into being without the other, though with enough concerted effort both may be brought into being together.

The sociology in question need not involve such macro-level thinking, instead focusing on concrete social actors. Latour describes the efforts of the engineer Rudolf Diesel to build an earlier new type of engine: "At the start, Diesel ties the fate of his engine to that of *any* fuel, thinking that they would all ignite at a very high pressure. . . . But then, nothing happened. Not every fuel ignited. This ally, which he had expected to be unproblematic and faithful, betrayed him. Only kerosene ignited, and then only erratically. . . . So what is happening? Diesel has to *shift his system of alliances*" (Latour 1987: 123; italics in the original). Diesel's alliances include entities as diverse as kerosene, pumps, other scientists and engineers, financiers and entrepreneurs, and possibly the consumer market. The technoscientist needs to remain constantly aware of a shifting array of dramatically different actors to succeed. A stable network, and a successful piece of technoscience, is the result of managing all of these actors so that they contribute toward a goal.

Actors build networks. These networks might resemble machines, when their components are made to act together to achieve a consistent effect. Or they might resemble facts, when their components are made to act as if they are in agreement. The work of technoscience is the work of understanding the interests of a variety of actors, and *translating* (both in place and in form) those interests so that the actors work together or in agreement (Callon 1986; Callon and Law 1989).

ANT is a materialist theory. Science and technology work by translating material actions and forces from one form into another. Scientific representations are the result of material manipulations, and they are solid precisely to the extent that they are mechanized. The rigidity of translations is key here. Data, for example, is valued as a form of representation because it is supposed to be the direct result of interactions with the natural world. Visiting an ecological field site in Brazil, Latour (1999) observes researchers creating data on the colors of soil samples. Munsell color charts are held against the samples (just as a painter will hold a color chart against a paint sample) so that the color of the sample can be translated into a uniform code. As Latour jokes, the gap between representation and the world, a way of seeing a standard philosophical problem, is reduced by scientists to a few millimeters. Data-level representations are themselves juxtaposed to form new relationships that are summarized and otherwise manipulated to form higher-level representations, representations that are more general and further from their objects. Again, the translation metaphor is apt, because these operations can be seen as translations of representations into new forms, in which they will be more generally applicable. Ideally, there are no leaps between data and theory –

and between theory and application – but only a series of minute steps. Thus there is no "action at a distance," though there may be long-distance control (see Star 1989).

That there is no action at a distance is a methodological as well as an empirical claim (Latour 1983). The working of abstract theories and other general knowledge appears a miracle unless it can be systematically traced back to local interactions, via hands-on manipulation and working machines, via data, and via techniques for summarizing, grouping, and otherwise exploiting information. Therefore, science and technology must work by translating material actions and forces from one form into another. Universal scientific knowledge is the product of the manipulation of local accounts, a product that can be transported to a wide variety of new local circumstances. It is only applicable through a new set of manipulations that adapt it once again to those local circumstances (or adapt those local circumstances to it).

Seen in these terms, laboratories give scientists and engineers power that other people do not have, for "it is in the laboratories that most new sources of power are generated" (Latour 1983: 160). The laboratory contains tools, like microscopes and telescopes, that change the effective sizes of things. Such tools make objects human in scale, and hence easier to study. The laboratory also contains a seemingly endless variety of tools for separating parts of objects, for controlling them, and for subjecting them to tests: objects are tested to find out what they can and cannot do. This process can also be thought of as a series of tests of actors, to find out which alliances can and cannot be built. Simple tools like centrifuges, vacuum pumps, furnaces, and scales have populated laboratories for hundreds of years; these and their modern descendants tease apart, stabilize, and then quantify objects (Carroll-Burke 2001). *Inscription devices*, or machines that "transform pieces of matter into written documents" (Latour and Woolgar 1986: 51), allow the scientist to deal with nature on pieces of paper. Like the representations produced by telescopes and microscopes these are also medium-sized, but perhaps more importantly they are durable, transportable, and relatively easy to compare to each other. Such *immutable mobiles* can be circulated and manipulated independently of the contexts from which they derive. Nature brought to a human scale, teased into components and made stable in the laboratory or other *center of calculation*, and turned into marks on paper or in a computer, is manipulable.

We can see that, while ANT is a general theory, it is one that explains the centrality of science and technology to the idea of modernity (Latour 1993b). Science and technology explicitly engage in crossing back and forth between objects and representations, creating situations in which humans and non-humans affect each other.

Box 7.1   The Pasteurization of France

Louis Pasteur is perhaps the model scientist for Latour. Pasteur's an-thrax vaccine is the subject of an early statement of ANT (Latour 1983), and his broader campaign on the microbial theory of disease is the subject of a short book (Latour 1988). In the spirit of Tolstoy's *War and Peace*, Latour asks how Pasteur could be seen as the central cause of a revolution in medicine and public health, even though he, as a single actor, can do almost nothing by himself.

The laboratory is probably the most important starting point. Here is Pasteur, describing the power of the laboratory:

> As soon as the physicist and chemist leave their laboratories, . . . they become incapable of the slightest discovery. The boldest conceptions, the most legitimate speculations, take on body and soul only when they are consecrated by observation and experience. Laboratory and discov-ery are correlative terms. Eliminate the laboratories and the physical sciences will become the image of sterility and death. . . . Outside the laboratories, the physicists and chemists are unarmed soldiers in the battlefield.   (in Latour 1988: 73)

Pasteur uses the strengths of the laboratory to get microbes to do what he wants. Whereas in nature microbes hide, being invisible components of messy constellations, in the laboratory they can be isolated and nur-tured, allowing Pasteur and his assistants to deal with visible colonies. These can thus be tested, or subjected to *trials of strength*, to find their properties. In the case of microbes, Pasteur is particularly interested in finding weak versions that can serve as vaccines.

Out of the complex set of symptoms and circumstances that is a disease, Pasteur *defines* a microbe in the laboratory; his manipula-tions and records specify its boundaries and properties. He then is able to argue, to the wider scientific and medical community, that his microbe is responsible for the disease. This is in part done via public demonstrations that repeat laboratory experiments – breakthroughs like the successful vaccination of sheep against anthrax are carefully staged demonstrations, in which the field is turned into a laboratory, and the public is invited to witness the outcomes of already-performed experiments. Public demonstrations help to convince people of two important things: that microbes are key to their goals, whether those goals are health, the strength of armies, or public order; and that Pasteur has control over those microbes.

Microbes can be seen as not merely entities that Pasteur studies, but agents with whom Pasteur builds an alliance. The alliance is ultimately very successful, creating enormous interest in Pasteur's methods of inquiry, reshaping public health measures, and gaining prestige and power for Pasteur. Thus we might see Pasteur's work as introducing a new element into society, an element of which other people have to take account if they are to achieve their goals.

If he is successful in all of these steps, then it should be easy for Pasteur to convince people to act around his conceptions of disease and health. When doctors, hygienists, regulators, and others put in place measures oriented around his purified microbe, it becomes a taken-for-granted truth that the microbe is the real cause of the disease, and that Pasteur is the cause of a revolution in medicine and public health.

While ANT is thoroughly materialist, it is also built on a relational ontology; it is based on a *relational materiality* (Law 1999). Objects are defined by their places in networks, and their properties appear in the context of tests, not in isolation. Perhaps most prominently, not only technoscientific objects but also social groups are products of network-building. Social interests are not fixed and internal to actors, but are changeable external objects. The French military of 1880 might be interested in recruiting better soldiers, but Pasteur translates that interest, via some simple rhetorical work, into support for his program of research. After Pasteur's work, the military has a new interest in basic research on microbes. Translation in ANT's sense is not neutral, but changes interests.

Whereas the strong programme was "symmetric" in its analysis of truth and falsity and in its application of the same social explanation for, say, both true and false beliefs, ANT is "supersymmetric," treating both the social and material worlds as the products of networks (Callon and Latour 1992; Callon and Law 1995). Representing both human and non-human actors, and treating them in the same relational terms, is one way of prompting full analyses, analyses that do not discriminate against any part of the ecologies of scientific facts and technological objects. It does not privilege any particular set of variables, because every variable (or set of actors) depends upon others. Networks confront each other as wholes, and to understand their successes and failures science and technology studies (S&TS) has to study the wholes of those networks.

Box 7.2   Ecological thinking

Science and technology are done in rich contexts that include material circumstances, social ties, established practices, and bodies of knowledge. Scientific and technological work is performed in complex ecological circumstances; to be successful that work must fit into or reshape its environment.

An ecological approach to the study of science and technology emphasizes that multiple and varying elements contribute to the success of an idea or artifact – and any element in an idea or artifact's environment may be responsible for failure. An idea does not by itself solve a problem, but needs to be combined with time to develop it, skilled work to provide evidence for it, rhetorical work to make it plausible to others, and the support to put all of those in place. If some of the evidential work is empirical, then it will also demand materials, and the tinkering to make the materials behave properly. Solutions to problems, therefore, need nurturing to succeed.

There is no *a priori* ordering of such elements. That is, no one of them is crucial in advance. With enough effort, and with enough willingness to make changes elsewhere in the environment, anything can be changed, moved, or made irrelevant. As a result, there is no *a priori* definition of good and bad ideas, good and bad technologies. Success stories are built out of many distinct elements. They are typically the result of many different innovations, some of which might normally be considered technical, some economic, some social, and some political. The "niche" of a technological artifact or a scientific fact is a multi-dimensional development.

Some Objections to Actor-Network Theory

Actor-network theory, especially in the form articulated by Bruno Latour in his widely read book *Science in Action* (1987), has become a constant touchstone in S&TS, and is increasingly being imported into other domains. The theory is easy to apply to, and productive of insights in, an apparently limitless number of cases. Its insistent focus on the materiality of relations creates research problems that can be solved. And its claims about the relationality of materials mean that its applications are often counter-intuitive. But this success does not mean that S&TS has uncritically accepted actor-network theory. The remainder of this chapter is devoted to criticisms of the theory. This discussion of the problems that ANT faces is not supposed to indicate the theory's failure. Rather, it contributes to an explanation of the theory, and to demonstrating its scope.

1   Practices and cultures

Actor-network theory, and for that matter almost every other approach in S&TS, portrays science as rational in a means–end sense: scientists use the resources that are available – rhetorical resources, established power, facts, and machines – to achieve their goals. Rational choices are not made in a vacuum, or even only in a field of simple material and conceptual resources. They are made in the context of existing cultures and practices of science and technology. Practices can be thought of as the accepted patterns of action and styles of work; cultures `define the scope of available resources (Pickering 1992a). Opportunistic science, even science that transforms cultures and practices, is an attempt to appropriately combine and recombine cultural resources to achieve scientists' goals. Practices and cultures provide the context and structure for scientific opportunism. But because ANT treats humans and non-humans on the same footing, and because it adopts an externalized view of actors, it does not pay attention to such distinctively human and apparently subjective factors as cultures and practices.

In his *Constructing Quarks*, for example, Andrew Pickering argues that scientists' judgments about which theories are most likely to be productive depend upon their own skills (Pickering 1984). Researchers will more highly value a theory the exploration of which demands skills, such as mathematical skills, that they already have or can easily obtain. At the same time, judgments about which skills are most likely to be productive will depend upon the theories that prevail, and thus choices about theories can redefine the culture and practices of the field.

Cultural networks do not fit neatly into the network framework offered by ANT. Trust is an essential feature of scientific and technological work, in that researchers rely upon findings and arguments made by people they have never met, and about whom they may know almost nothing. But trust is often established through faith in a common culture. Steven Shapin (1994) argues that the structure of trust in science was laid down by being transferred from the structure of gentlemanly trust in the seventeenth century; gentlemen could trust each other, and could not easily challenge each other's truthfulness. Similarly, trust in technical judgment often resides in cultural affiliations. Engineers educated in the École Polytechnique in France trust each other's judgments (Porter 1995), just as do engineers educated at the Massachusetts Institute of Technology (e.g. MacKenzie 1990).

Therefore, either practices and cultures are themselves actors, which is counter-intuitive at the least, or something has to be added to the culturally flat world of ANT to account for rational choices.

2   Problems of agency

Actor-network theory has been criticized for its distribution of agency. On the one hand, it may encourage analyses centered on key figures; many of Latour's examples are of heroic scientists and engineers, or of failed heroes. Such centering may make the world appear to revolve around these heroes or near-heroes. The stories that result miss work being done by other actors, miss structures that prevent others from participating, and miss non-central perspectives. Marginal, and particularly marginalized, perspectives may provide dramatically different insights; for example, women who are sidelined from scientific or technical work may see the activities of science and technology quite differently (see, e.g., Star 1991). With ANT's focus on agency, positions from which it is difficult to act make for less interesting positions from which to tell stories. So ANT may encourage the following of heroes and would-be heroes.

On the other hand, actor-network analyses can be centered on any perspective, or on multiple perspectives. Michel Callon even famously uses the perspective of the scallops of St. Brieuc Bay for a portion of one statement of ANT (Callon 1986). This positing of non-human agents is one of the more controversial features of the theory, attracting a great deal of criticism (see, e.g., Collins and Yearley 1992).

In principle, ANT is entirely symmetrical around the human/non-human divide. Given its externalized perspective, non-humans can appear to act in exactly the same way as do humans – they can have interests, they can enroll others. (Strictly speaking, all of the actors of ANT are *actants*, or things made to act. Thus agency is an effect of networks, not prior to them. This is a difficult distinction to sustain, and the ends of ANT's analyses seem to rest on the agency of non-humans.) In practice, though, actor-network analyses tend to downplay any agency that non-humans might have (e.g. Miettinen 1998). Humans appear to have richer repertoires of strategies and goals than do non-humans, and so make more interesting subjects of study. The subtitle of Latour's popular *Science in Action* is *How to Follow Scientists and Engineers through Society*, indicating that however symmetric ANT is, of interest are the actions of scientists and engineers.

3 Problems of realism

Running parallel to problems of agency are problems of realism. On the one hand, ANT's relationalism would seem to turn everything into an outcome of network-building. Before their definition and public circulation, through laboratory and rhetorical work, natural objects cannot be said to have any real scientific properties. Before their public circulation and use, artifacts cannot be said to have any real technical properties, to do anything. For this reason, ANT is often seen as a blunt version of constructivism: what is, is constructed by networks of actors. This constructivism flies in the face of strong intuitions that scientists discover, rather than help create, the properties of natural things. It flies in the face of strong intuitions that technological ideas have or lack force of their own accord, whether or not they turn out to be successful. And this constructivism runs against the

arguments of realists that (at least some) things have real and intrinsic properties, no matter where in any network they sit.

On the other hand, positing non-human agents appears to commit ANT
to realism. Even if ANT assumes that scientists in some sense define or
construct the properties of the so-called natural world, it takes their interests seriously. That is, even if an object's interests can be manipulated, they
resist that manipulation, and hence push back against the network. This
type of picture assumes a reality that is prior to the work of scientists, engineers, and any other actors. Latour says, "A little bit of constructivism
takes you far away from realism; a complete constructivism brings you back
to it" (Latour 1990: 71).

The implicit realism of ANT has been both criticized, as a step backwards from the successes of methodological relativism (e.g. Collins and
Yearley 1992), and praised as a way of integrating the social and natural
world into S&TS (Sismondo 1996). For the purposes of this book, whether
ANT does make realist assumptions, and whether they might move the
discussion forwards or backwards are left as open questions, much as they
have been in the field itself.

4   Problems of the stability of objects and actions

The last problem facing ANT to be mentioned here is one that will be
made increasingly salient in later chapters. According to ANT, the power
of science and technology rests in the arrangement of actors so that they
form literal and metaphorical machines, combining and multiplying their
powers. That machining is possible because of the power of laboratories
and laboratory-like settings (such as field sites that are made to mimic labs;
see Latour 1999). And the power of laboratories depends upon relatively
formulaic observations and manipulations. Once an object has been defined and characterized, it can be trusted to behave similarly in all similar
situations, and actions can be delegated to that object.

Science and technology gain power from the translation of forces from
context to context, translations that can only be achieved by formal rules.
Colorful language aside, this picture overlaps with the picture put forward
by the logical positivists in the 1930s, who also saw the successes of science
and technology as only explicable in terms of formal rules. However, rules
have to be interpreted, and Wittgenstein's problem of rule-following shows
that no statement of a rule can determine its interpretations (box 3.2). As
we will see, S&TS has shown how much of the work of science and technology involves tinkering, how difficult the work of making observations
and manipulations is, how much expert judgment is involved in routine
science and engineering, and how that judgment is not reducible to formulas (chapter 9). ANT, while it recognizes the provisional and challenge-

able nature of laboratory work, glides over these issues. It presents science and technology as powerful because of the relative rigidity of their translations, or the objectivity – in the sense that they capture objects – of their procedures. Yet rigidity of translation may be a fiction, hiding many layers of expert judgment.

Conclusions

Since the publication of Latour's *Science in Action* (1987), ANT has dominated theoretical discussions in S&TS, and has served as a framework for an enormous number of studies. Its successes, as a theory of science, technology, and everything else, have been mostly bound up in its relational materialism. As a materialist theory it explains intuitively the successes and failures of facts and artifacts: they are the effects of the successful translation of actions, forces, and interests. As a relationalist theory it suggests novel results and promotes ecological analyses: humans and non-humans are bound up with each other, and features on neither side of that apparent divide can be understood without reference to features on the other. Whether actor-network theorists can answer all the questions people have of it remains to be seen, but it stands as the most successful of S&TS's theoretical achievements so far.

CHAPTER 8

Two Questions Concerning Technology

Is Technology Applied Science?

The idea that technology is applied science is now centuries old. In the early seventeenth century, Francis Bacon and René Descartes both argued for the value of scientific research by claiming that it would produce useful technology. In the twentieth century this view was championed most importantly by Vannevar Bush, one of the architects of the science policy pursued by the United States after the Second World War: "Basic research . . . creates the fund from which the practical applications of knowledge must be drawn. New products and new processes do not appear full-grown. They are founded on new principles and new conceptions, which in turn are painstakingly developed by research in the purest realms of science. . . . Today, it is truer than ever that basic research is the pacemaker of technological progress."

The view that technology is applied science has been challenged from many directions. In particular, accounts of artifacts and technologies show that scientific knowledge plays relatively little direct role in the development even of many state-of-the-art technologies. Historians and other theorists of technology have argued that there are technological knowledge traditions that are independent of scientific knowledge traditions, and that to understand the artifacts one needs to understand those knowledge traditions. At the same time, however, some people working in science and technology studies (S&TS) have argued that science and technology are not sufficiently well defined and distinct for there to be any determinate relationship between them.

Because of its large investment in basic research, in the mid-1960s the US Department of Defense conducted audits to discover how valuable that research was. Project Hindsight was a study of the key events leading to the development of 20 weapons systems. It classified 91 percent of the key events as technological, 8.7 percent as applied science, and 0.3 percent as basic science. Project Hindsight thus suggested that the direct influence of science on technology was very small, even within an institution that invested heavily in science, and which was at some key forefronts of technological development. A subsequent study, TRACES, challenged that picture by looking at prominent civilian technologies and following their origins further back in the historical record.

Among historians of technology it is widely accepted that "science owes more to the steam engine than the steam engine owes to science." For example, work on the history of aircraft suggests that aeronautical engineering is relatively divorced from science: engineers consult scientific results when they see a need to, but there is no sense in which their work is driven by science or in which it is the application of science (Vincenti 1990). Engineers develop their own mathematics, their own experimental results, and their own techniques. Or, similarly, the innovative electrical engineer Charles Steinmetz did not either apply physical theory or derive his own theoretical claims from it (Kline 1992), instead developing theoretical knowledge in purely engineering contexts. This is despite the fact that he considered his work to be applied science; it was applied science in that it fit Steinmetz's understanding of the scientific method.

Edwin Layton (1971, 1974) argues that the reason why the model of technology as applied science is so pervasive, and yet individual technologies cannot be seen to depend on basic science, is that technological knowledge is downplayed. Engineers and inventors participate in knowledge traditions, which shape the work that they do. Science, then, does not have a monopoly on technical knowledge. In the nineteenth century, for example, American engineers developed their own theoretical works on the strength of materials, drawing on but modifying earlier scientific research. When engineers needed results that bore on their practical problems, they looked to engineering research, not pure science.

The idea that there are important traditions of technological knowledge has resonated with and influenced the thinking of other historians and philosophers of technology. Rachel Laudan's problem-solving model of technology (1984) assumes that the development of technologies is a research process, driven by interesting problems. The sources of problems Laudan identifies are: actual and potential functional failure of current technologies, extrapolation from past technological successes, imbalances between related technologies, and, much more rarely, external needs demanding a technical solution. It is notable that all but the last of these problem sources stem from within technological knowledge traditions. Edward Constant's (1984) Kuhnian model of normal and revolutionary technology adopts a very similar picture of technology as research, and a similar picture of engineering and other technological knowledge forming its own traditions.

As we will see in later chapters, for a group of people to have its own tradition of knowledge suggests that that knowledge will be tied to the group's social networks and material circumstances. There is a measure of practical incommensurability between knowledge traditions, seen in difficulties of translation. In addition, some knowledge within a tradition is tacit, not fully formalizable, and requires socialization for it to be passed from person to person.

So far, we have seen arguments for the autonomy of technological practice. A separate set of arguments challenges the picture of technology as applied science by insisting on the lack of distinctness of science and technology. For example, in a study of the idea of large technological systems, the historian of technology Thomas Hughes claims that "persons committed emotionally and intellectually to problem solving associated with system creation and development rarely take note of disciplinary boundaries, unless bureaucracy has taken command" (Hughes 1987: 64). "Scientists" invent, and "inventors" do scientific research – whatever is necessary to push forward their systemic program.

Actor-network theory's term *technoscience* suggests a pragmatic characterization of scientific knowledge. For the pragmatist, scientific knowledge is about what natural objects can be made to do. Thus Knorr Cetina's book *The Manufacture of Knowledge* (1981) argues that laboratory science is about what can be constructed, not about what exists independently. Peter Dear (1995) argues that the key to the scientific revolution was a new orientation to experimental inquiry. Even though it produced unnatural objects, experimentation became acceptable because it showed what could be constructed of the natural world. Patrick Carroll-Burke (2001) argues that this practical orientation has remained intact, that science is defined by its use of a number of different types of "epistemic engines," devices for making natural objects more knowable. For the purposes of this chapter, the pragmatic orientation is relevant in that it draws attention to the ways in which science depends upon technology, both materially and conceptually.

The term *technoscience* also draws attention to the increasing interdependence of science and technology. We might see it as odd that historians are insisting on the autonomy of technological traditions and cultures precisely when there is a new spate of science-based technologies and technologically-oriented science – biotechnologies, new materials science, and nanotechnology all cross obvious lines.

Latour's networks and Hughes's technological systems bundle many different resources together. Thomas Edison freely mixed economic calculations, the properties of materials, and sociological concerns in his designs (Hughes 1985). Technologists need scientific and technical knowledge, but they also need material, financial, social, and rhetorical resources. Even ideology can be an input, in the sense that it might shape decisions and the conditions of success and failure (e.g. Kaiserfeld 1996). For network builders nothing can be reduced to only one dimension. Technology requires heterogeneous engineering of a dramatic diversity of elements (Law; 1987; Bucciarelli 1994). A better picture of technology, then, is one that incorporates many different inputs, rather than being particularly dependent upon a single stream. It is possible that no one input is even essential, and could not be worked around, given enough hard work, ingenuity and other resources.

To sum up, scientific knowledge is one resource on which engineers and inventors can draw, and perhaps on which they are drawing increasingly. But there is no reason to see it as a dominant resource. Rather, the development of technology is a complex process that integrates many different things: different kinds of knowledge – including its own knowledge traditions – and different kinds of material resources. At the same time, it is clear that science draws on technology for its instruments, and perhaps also for some of its models of knowledge, just as some engineers may draw on science for their models of engineering knowledge. There are multiple relations of science and technology, rather than a single monolithic relation.

Box 8.1   Changing modes of research

Whatever the relations are between science and technology, they are shifting. For example, although corporate research has always been supposed to lead to profits, many companies created research units the goals of which were shielded from business concerns so that they could pursue open-ended questions. Since the 1980s, restructuring of many high-technology companies especially in the United States has led to the shrinking of those laboratories, and to their changing relationships with the rest of the company. One researcher comments that "I have jumped from theoretical physics to what customers would like the most" (Varmi 2000: 405). In part this is a change in the notion of the purposes of science, but it may also represent a change in people's sense of science's technological potential – as the idea of the "knowledge economy" suggests.

Going alongside this is a sense that universities, and academic research, are also changing. Universities, and university researchers, are increasingly patenting their results, and entering into partnerships with corporations to fund research and develop products. There have been a number of different formulations of the changing structures of research. Much discussed is the idea that there is a new "mode" of knowledge production (Gibbons et al. 1994; Ziman 1994). Instead of the discipline-bound problem-oriented research of mode 1, mode 2 involves transdisciplinary work, possibly involving actors from a variety of types of organizations, in which the application is clearly in view. Some of the same authors as put forward the mode 2 concept have developed an alternative that sees an increase in "contextualization" of science, a process in which non-scientists become involved in shaping the direction and content of specific pieces of research (Nowotny, Scott, and Gibbons 2001). There might be an increase in the importance of research that is simultaneously basic and applicable (Stokes 1997). Another alternative formulation of the change is in the more organizational terms of a "triple helix" of university–industry–government in interaction (Etzkowitz and Leydesdorff 1997). Governments are demanding of universities that they be relevant in trade for support, universities are becoming entrepreneurial, and industry is buying research from universities. The change might also be put in more negative terms, that there is a crisis in key categories that support the idea of pure scientific research: the justification for and bounds of academic freedom, the public domain, and disinterestedness have all become unclear, disrupting the ethos of pure science (McSherry 2001).

Critics of each of these formulations charge that the changes described are not nearly as abrupt as they are portrayed to be, but these criticisms do not take away from the feeling that dramatic changes are under way for scientific research, and that these changes are connected to the potential application of that research. Whether or not there is less pure science, there is certainly more applied science. Science and technology are seen as increasingly connected.

Does Technology Drive History?

A few of Karl Marx and Friedrich Engels's memorable comments on the influence of technology on economics and society can stand in for the position of the technological determinist, though they are certainly not everything that Marx and Engels had to say about the determinants of social structures. Looking at large-scale structures, Marx famously said: "The hand-mill gives you society with the feudal lord, the steam-mill, society with the industrial capitalist." Engels, talking about smaller-scale structures, claimed that "the automatic machinery of a big factory is much more despotic than the small capitalists who employ workers ever have been."

Technological determinism is the position that material forces, and especially the properties of available technologies, determine social events. The reasoning behind it is usually economistic: the available material resources, and the technologies for manipulating those resources, form the environment in which rational economic choices are made. In addition, technological determinism emphasizes "real-world constraints" and "technical logics" that shape technological trajectories (Vincenti 1995). Therefore, social variables ultimately depend upon material ones.

There are a number of different technological determinisms (see Bimber 1994), but the central idea is that technological changes force social adaptations, and through this narrowly constrain the trajectories of human history. Robert Heilbroner, supporting Marx, says that

> the hand-mill (if we may take this as referring to late medieval technology in general) required a work force composed of skilled or semiskilled craftsmen, who were free to practice their occupations at home or in a small atelier, at times and seasons that varied considerably. By way of contrast, the steam-mill – that is, the technology of the nineteenth century – required a work force composed of semiskilled or unskilled operatives who could work only at the factory site and only at [a] strict time schedule.   (Heilbroner 1994 [1967]: 60)

Because economic actors make rational choices, class structure is determined by the dominant technologies. This sort of reasoning applies to the largest scales, but also to much more local decisions. Thus technology shapes economic choices, and through those shapes history.

Some technologies appear particularly compatible with some types of political and social arrangements. In a well-known essay, Langdon Winner (1986a) asks: "do artifacts have politics?" He concludes that they do. Following Engels, he argues that some complex technological decisions will lend themselves to more hierarchical organization than others, in the name of efficiency – the complexity of modern industrial production does not lend itself to consensus decision-making. In addition, Winner argues, some technologies, such as nuclear power, are dangerous enough that they may bring their own demands for policing, and other forms of state power. And finally, individual artifacts may be constructed to achieve political goals. In a much-cited example, Winner describes how New York's "Master Builder" of roads, Robert Moses, designed overpasses on Long Island's parkways that were low enough to discourage buses, thus reserving Long Island's beaches for the car-owning classes. This example has been shown to be less straightforward than Winner originally portrayed it (Joerges 1999), but the point can be made with mundane examples. Speed bumps perform the political purpose of reducing and slowing traffic on a street, and thus increasing the property values of a family-oriented neighborhood; this technology becomes a form of long-distance control (Latour 1993a). To take a different type of example, David Noble argues that in the history of industrial automation choices were made to disempower key groups (Noble 1984). Numerical control automation, the dominant form, was developed to eliminate machinist skill altogether from the factory floor, and therefore to eliminate the power of key unions. While also intended to reduce factories' dependence on skilled labor, record-playback automation, a technology not developed nearly as much, would have required the maintenance of machinist skill, to reproduce it in machine form (for some related issues, see Wood 1982).

Even for non-determinists, the effects of technologies can form an important site of study. As we saw in chapter 1, a key part of the pre-S&TS constellation of ideas on science and technology was the study of the positive and negative effects of technologies, and the attempt to think systematically about these effects. That type of work continues, and is part of the more practical and applied portion of S&TS. At the same time, some researchers have been challenging a seemingly unchallengeable assumption, the assumption that technologies have any systematic effects at all! In fact, they challenge something slightly deeper, the idea that technologies have essential features. If technologies have no essential features, then they should not have systematic effects, and if they do not have any systematic effects then they cannot determine the structure of the social world.

No technology – and in fact no object – has only one potential use. Even something as apparently purposeful as a watch can be simultaneously constructed to tell time, to be attractive, to make profits, to refer to a well-known style of clock, to make a statement about its wearer, etc. Even the apparently simple goal of telling time might be seen as a multitude of different goals: within a day one might use a watch to keep on schedule, to find out how long a bicycle ride took, to regulate the cooking of a pastry, to notice when the sun set, and so on. Given this diversity, there is no essence to a watch. And if the watch has no essence, then we can only say that it has systematic effects within a particular human environment. Change that environment and one changes what the watch does.

Trevor Pinch and Wiebe Bijker (1987), in their work on "Social Construction of Technology" (SCOT), develop this point into a framework for thinking about the development of technologies. In their central example, the development of the safety bicycle, the basic design of most twentieth-century bicycles, there is an appearance of inevitability about the outcome. The standard modern bicycle is stable, safe, efficient, and fast, and therefore we might see its predecessors as important, but ultimately doomed, steps toward the safety bicycle. On Pinch and Bijker's analysis, though, the safety bicycle did not triumph because of an intrinsically superior design. Some users felt that other early bicycle variants represented superior designs, at least superior to the early versions of the safety bicycle with which they competed. For many young male riders, the safety bicycle sacrificed style for a claim to stability, even though new riders did not, of course, find it very stable. Young male riders were one *relevant social group* that was not appeased by the new design. There is *interpretive flexibility* both in the understanding of technologies and in their design. We should see trajectories of technologies as the result of rhetorical operations, defining the users of artifacts, their uses, and the problems that particular designs solve.

On a SCOT analysis, the success of an artifact depends in large part upon the strength and size of the group that takes it up and promotes it. Its definition depends upon the associations that different actors make. Interpretive flexibility is thus a necessary feature of artifacts, because what an artifact does and how well it performs are the results of a competition of different groups' claims. Thus the good design of the safety bicycle cannot be the mover behind its success; good design is instead the result of its success.

Keith Grint and Steve Woolgar (1997) have developed this argument further. They take the metaphor of "technology as text," and try to show that technologies are interpretable as flexibly as are texts. They show, for example, that the Luddites of early nineteenth-century Britain adopted a variety of interpretations of the factory machines that they did and did not smash. Although some saw the factory machines as upsetting their preferred modes of work, others saw the problem in the masters of the factories. And, they argue, resistance to the new technologies diminished when new left-wing political theories articulated the machines as saviors of the working classes. The machines, then, did not have any single set of effects.

A study done by Woolgar shows some of the work needed to give technologies determinate meanings. Woolgar acted as participant-observer in a computer firm that was in the process of developing a computer, with bundled software, for educational markets. At a point well into the development process, the firm needed to test prototypes of their package, to see how easy it was for unskilled users to figure out how to perform some standard tasks. On the one hand these tests could be seen as revealing what needed to be done to the computers in order for them to be more user-friendly. On the other hand they could be seen as revealing what needed to be done to the users – how they needed to be defined, educated, controlled – to make them more computer-friendly: successful technologies require *configuring the user*. The computer, then, does what it does only in the context of an appropriate set of users.

But surely some features of technologies defy interpretation? We might, for example, ask with Rob Kling (1992): "What's so social about being shot?" Everything, say Grint and Woolgar. In a tour de force of anti-essentialist argumentation, Grint and Woolgar argue that a gun being shot is not nearly as simple a thing as it might seem. It is clear that the act of shooting a gun is intensely meaningful – some guns are, for example, more manly than others. But more than that, even injuries by gunshot can take on different meanings. When female Israeli soldiers were shot in 1948, "men who might have found the wounding of a male colleague comparatively tolerable were shocked by the injury of a woman, and the mission tended to get forgotten in a general scramble to ensure that she received medical aid" (R. Holmes, quoted in Grint and Woolgar 1997: 159). Even death is not so certain. Leaving aside common uncertainties about causes of death and the timing of death, there are cross-cultural differences about death and what happens following it. No matter how unmalleable a technology might look, there are always situations, some of them highly theoretical, in which the technology can take on unusual uses or interpretations.

To accept that technologies do not have essences is to pull the rug out from under technological determinism. If they do nothing outside of the social and material contexts in which they are developed and used, technologies cannot be the real drivers of history. Rather these contexts are in the drivers' seats. This recognition is potentially useful for political analyses, as particular technologies can be used to affect social relations (Hård 1993). This has been explored most in the context of labor relations (e.g. Noble 1984) and in the context of gender relations (e.g. Cockburn 1985; Cowan 1983; Kirkup and Smith Keller 1992).

There should, then, be no debate about technological determinism. However, in practice nobody holds a determinism that is strict enough to be completely overturned by these arguments. Even the strictest of determinists, like Heilbroner, admit that social forces play a variety of important roles in producing and shaping technology's effects. Such a "soft" determinism is an interpretive stance, according to Heilbroner, that directs us to look first to technological change to understand economic change. Choices are made in relation to material resources and opportunities. To the extent that we can see social choices as economic choices, technology will play a key role.

The anti-essentialist strain that has developed within S&TS is a counterbalancing interpretive stance. Anti-essentialists show us that even soft determinism must be understood within a social framework, in that the properties of technologies can be determinative of social events only once the social world has established what the properties of technologies are. It thus directs us to look to the social world to understand technological change and its effects. This is perhaps most valuable for its constant reminder that things could be different.

Just paying attention to anti-essentialist lessons would remove the interest in studying technology. We study technology because artifacts appear to do things, or at least are made to do things. Therefore we need a non-deterministic theory of technology that can accommodate the obduracy of artifacts. We need something like Wiebe Bijker's theory of *sociotechnology*, a theory of the thorough intertwining of the social and technical (Bijker 1995). This theory draws heavily on work on technology as heterogeneous engineering, and on Bijker's work with Trevor Pinch on SCOT. A key concept in Bijker's theory is that of the *technological frame*, the set of practices and the material and social infrastructure built up around an artifact or collection of similar artifacts (Bijker 1995: 123). As the frame is developed, it guides future actions. A technological frame, then, may reflect engineers' understandings of the key problems of the artifact, and the directions in which solutions should be sought. It may also reflect understandings of the potential users of the artifact, and users' understanding of its functions. If a strong technological frame has developed, it will cramp interpretive flexibility. The concept is therefore useful in helping to understand how technologies can appear deterministic, while only appearing so in particular contexts.

Even if technologies do not have essential forms – properties that they have independent of their interpretations, or functions that they perform independently of what they are made to do – essences can return, in muted form, as dispositions. For Francis Bacon, with whom we started this chapter, the reason that scientific research should translate into technological benefits was that scientific research investigated what substances could be made to do. The form of a substance, for Bacon, is its response to circumstances. The forms of substances are those underlying natures that express the potentialities of those substances. This is why Bacon says that "knowledge and human power are synonymous."

One might make the same argument about the properties of technological artifacts. Artifacts do nothing by themselves, though they can be said to have effects in particular social contexts. To the extent that we can specify the relevant features of their social contexts, we might say that technological artifacts have Baconian forms. Particular pieces of technology can be said to have definitive properties, though they change depending upon context. Material reductionism, then, only makes sense in a given social context, just as social reductionism only makes sense in a given material context.

Does technology drive history, then? History could be almost nothing without it. As Bijker puts it, "purely social relations are to be found only in the imaginations of sociologists or among baboons." But, equally, technology could be almost nothing without history. Bijker continues, "and purely technical relations are to be found only in the wilder reaches of science fiction" (Bijker 1995).

Box 8.2   Were electric automobiles doomed to fail?

David Kirsch's history of the electric vehicle illustrates both the difficulty and power of deterministic thinking (Kirsch 2000). The standard history of the internal combustion automobile portrays the electric vehicle as doomed to failure. Compared with the gasoline-powered vehicle, the electric vehicle suffered from lack of power and lack of range, and therefore could never be the all-purpose vehicle that consumers wanted. Kirsch argues, however, that "technological superiority was ultimately located in the hearts and minds of engineers, consumers, and drivers, not programmed inexorably into the chemical bonds of refined petroleum" (Kirsch 2000: 4).

Until about 1915 electric cars and trucks could compete with gasoline-powered cars and trucks in a number of market niches. In many ways electric cars and trucks were the more natural successors to horse-drawn carriages. Gasoline-powered trucks were faster than electric trucks, but for the owners of delivery companies speed was more likely to damage goods, to damage the vehicles themselves, and anyway was effectively limited in cities. Because they were easy to restart, electric trucks were better suited to making deliveries than early gasoline-powered trucks; this was especially true given the horse-paced rhythm of existing delivery service, which demanded interaction between driver and customer. Electric taxis were fashionable, comfortable, and quiet, and for a time were successful in a number of American cities, so much so that in 1900 the Electric Vehicle Company was the largest manufacturer of automobiles in the United States.

As innovators, electric taxi services were burdened with early equipment, they sometimes suffered poor management, and they were hit by expensive strikes. They also failed to participate in an integrated urban transit system that linked rail and road, to create a niche that they could dominate and in which they could then innovate. Meanwhile, Henry Ford's grand experiment in producing low-cost vehicles on assembly lines helped to spell the end of the electric vehicle. The First World War created a huge demand for gasoline-powered vehicles, which were better suited to war conditions than were electric ones. Increasing suburbanization of US cities meant that electric cars and trucks were restricted to a smaller and smaller segment of the market. Of course, that suburbanization was helped along by the successes of gasoline, and thus we might argue that the demands of consumers not only shaped, but were shaped by the technologies.

In 1900, then, the fate of the electric vehicle was not sealed. Does this failure of technological determinism mean that electric cars could be rehabilitated? According to Kirsch, that is unlikely. In 1900, both gasoline and electric cars fit the material and social contexts they faced, albeit very imperfectly. In the year 2000, material and social contexts have been shaped around the internal combustion engine. Given how unlikely it is that electric cars could compete directly with gasoline cars in these new contexts, their future does not look bright. Thus while a global technological determinism fails, within some contexts a soft determinism serves a useful heuristic purpose.

CHAPTER 9

Laboratories

The Idea of the Laboratory Study

In the 1970s a number of science and technology studies (S&TS) research-ers more or less simultaneously identified a novel approach to the study of science and technology. They went into laboratories to study the practical and day-to-day scientific work. Most prominent among the new students of laboratories were Bruno Latour and Steve Woolgar (Latour and Woolgar 1986 [1979]), Karin Knorr Cetina (1981), Harry Collins (1991 [1985]), Michael Lynch (1985), Sharon Traweek (1988), and Michael Zenzen and Sal Restivo (1982). At the same time, June Goodfield (1981) and Tracey Kidder (1981) published more novelistic ethnographies of science and tech-nology that arrived at some similar findings. Laboratories are exemplary sites because experimental work is a central part of scientific activity, and experimental work is relatively visible (though Michael Lynch (1991) points out that much work in the laboratory is not visible – researchers may sit at microscopes for hours at a time).

Many of the first group of students of laboratories used their observa-tions to make philosophical arguments about the nature of scientific knowl-edge, but framed their results anthropologically. In their well-known book *Laboratory Life*, Latour and Woolgar announce their intention to treat the scientists being studied as an alien tribe:

> Since the turn of the century, scores of men and women have penetrated deep forests, lived in hostile climates, and weathered hostility, boredom, and disease in order to gather the remnants of so-called primitive societies. By contrast to the frequency of these anthropological excursions, relatively few attempts have been made to penetrate the intimacy of life among tribes which are much nearer at hand. This is perhaps surprising in view of the reception and importance attached to their product in modern civilised societies: we refer, of course, to tribes of scientists and to their production of science. (Latour and Woolgar 1986 [1979]: 17)

Treating scientists as largely alien gives the analysts leave to see their practices as unfamiliar, and hence to ask questions of them that would not normally be asked. Like the methodological tenets of the strong programme, an anthropological approach puts a spotlight on the underdetermination of practices and facts. The task of the analyst becomes one of showing how their subjects construct their choices as determined.

Learning to See

One of the first things ethnographers found when visiting laboratories was that they, as outsiders, did not see what their informants saw. The very things that researchers took as data were often difficult to recognize as distinct objects or clear readings. That suggests an interesting site of study: experts have somehow learned how to read their material in a way that novices cannot.

Pictures of such things as electrophoresis gels of DNA fragments or radioactively tagged proteins do not provide unproblematic data: there is simultaneously too much information there, and not enough. Experts may see the patterns that they are looking for, perhaps after some puzzling and discussion, by making the pictures contain exactly the right amount of information (Zenzen and Restivo 1982; Charlesworth et al. 1989: 162–3). At the same time, the fact that experts can see what is relevant suggests that they may miss other interesting features of their material (Hanson 1958; Kuhn 1970a). They have to block out irrelevant material and emphasize relevant material, turning the picture into a diagram in their mind's eye. Here are Zenzen and Restivo describing the process:

> Kornbrekke's first problem was to determine what emulsion type he had after shaking. This sounds innocent enough, but it is fraught with difficulty. After shaking, most of the mixtures separate quite rapidly and one sees a colorless solution with a highly active boundary or interface between the two volumes of liquids. All the standard techniques for determining emulsion type or morphology apply only to *stable* emulsions. Kornbrekke had to teach himself to see what a short-lived emulsion "looks like". By "thinking about what you have" (in Kornbrekke's words) and hypothesizing relevant visual parameters, Kornbrekke had to make the data manifest themselves.   (Zenzen and Restivo 1982: 453)

Tinkering, Skills, and Tacit Knowledge

As Ian Hacking (1983) argues, laboratory work is not merely about representation, but about intervention: researchers are actively engaged in manipulating their materials. We have already seen something of the messiness of the laboratory. The messiness of observation finds its counterpart in the intervention side of laboratory life. A number of people who have looked closely at laboratory work have drawn attention to how often manipulations fail, apparatuses do not work, and materials do not behave. Hacking puts it bluntly, "experiments don't work" (Hacking 1983: 229). Laboratory research thus involves a tremendous amount of *tinkering* (Knorr Cetina 1981) or *bricolage* (Latour and Woolgar 1986 [1979]) in order to make recalcitrant material do what it is supposed to. Rational planning only covers a portion of the situations faced in the laboratory, and thus needs to be supplemented by skilled indexical reasoning, tied to particular problems.

H. M. Collins's studies of laboratories have provided several important concepts for understanding laboratories, and technical work more generally. The bulk of this chapter is given over to two key concepts, *tacit knowledge*, and *experimenters' regress*. Collins's work is grounded in *methodological relativism*, a notion parallel to the strong programme's strictures of symmetry and impartiality (see chapter 5). Methodological relativism is the assumption that there is no fact of the matter in controversies, an assumption that allows researchers to better study the processes by which those controversies are resolved.

Tacit knowledge

The idea of tacit knowledge is due to the chemist and philosopher Michael Polanyi (1958). Collins first puts it to use in his study of attempts by researchers to build a Transversely Excited Atmospheric Laser, or TEA-laser (Collins 1974, 1991 [1985]). In his analysis of the study, Collins established a number of points of reference that are still used to discuss skill in S&TS. In the early 1970s, he interviewed researchers at six different British laboratories, all of which were trying to build versions of a TEA-laser, and interviewed scientists at five North American laboratories from which the British groups had received information. The main attraction of the TEA-laser, which had been developed in a Canadian defense laboratory, was the low cost and robustness of its gas chamber. Most of the components could be easily purchased, and the design was not thought to be particularly complicated.

It turned out, however, that the transfer of knowledge about how to build the laser was difficult. Collins's results are striking: Nobody "succeeded in building a laser by using only information found in published or other written sources. . . . no scientist succeeded where their informant was a 'middle man' who had not built a device himself . . . even where the informant had built a successful device, and where the information flowed freely as far as could be seen, the learner would be unlikely to succeed without some extended period of contact with the informant" (Collins 1991: 55). According to Collins, this was not because of secrecy, as he observed cases in which there was no apparent secrecy, but in which knowledge did not travel easily. Rather, the transfer of knowledge about how to build the laser required more than simply a set of instructions; it required the passing on of a skill. Collins thus compares two models for the transfer of knowledge: the "algorithmic" model assumes that a set of formal instructions can suffice; the "enculturational" model assumes that socialization is necessary.

While some knowledge can be easily communicated, some resists formalization. Some relevant knowledge about TEA-lasers could be turned into information that could be written down and distributed, but some information could be communicated only through a socialization process. This is Polanyi's tacit knowledge. Tacit knowledge can be embodied knowledge, like that of how to ride a bicycle, how to grow a crystal, or how to clone an antibody. It can also be knowledge that is embedded in material and intellectual contexts, being dependent upon particular material and intellectual arrangements for its success and even meaningfulness. That scientific knowledge has a kind of material and intellectual locality is seen in almost every study in S&TS. Collins has devoted a considerable amount of effort, and several books (Collins 1990; Collins and Kusch 1998) to arguing that some knowledge is in principle tacit, when it is irreducibly tied to social actions. This amounts to saying that artificial intelligence can never fully replace expertise, because it could not be formalized (box 9.1).

There is another crucial aspect of scientific and technical skill that we can pick up from the TEA-laser study. In addition to interviews, Collins spent time in laboratories observing and helping with the construction of lasers. That allowed him to see some of the difficulties first-hand. One of his subjects, a Dr. Bob Harrison at the University of Bath, tried to make at least two of them for use in his lab. Both lasers were difficult to get in working order, even though Harrison apparently had all of the needed technical knowledge and expertise. There were details about almost every component and its placing that potentially caused trouble, and obvious frustration:

> I was then in a situation of – well, what shall I do with this damn laser? It's arcing and I can't [monitor the discharge pulse] so I thought well, let's take a trip to [the contact laboratory] and just make sure that simple characteristics like length of leads, glass tubes . . . look right   (Harrison; quoted in Collins 1991: 61)

The first laser owed its working to considerable advice from Harrison's informants. But the second laser was equally difficult, even though it was a near copy of the first! In complex contexts, then, technical skills are capricious. Judgments about what matters have to be developed on the basis of considerable experience, but much science and technology operates in relatively novel contexts – dealing with new, contentious or poorly understood phenomena, or dealing with new, untested, or poorly understood techniques – and about which there is little experience. So at least on the forefront of research and development, many scientific and technical skills leave considerable room for uncertainty.

Box 9.1   A consequence: limitations of artificial intelligence

Hubert Dreyfus's 1972 book, *What Computers Can't Do*, argues that there are some domains in which artificial intelligence (AI) is impossible if the AI is based on programmed-in expertise. This provocative claim flies in the face of the promises made by researchers on and promoters of AI since the invention of computers. Dreyfus's argument turns on a particular notion of tacit knowledge: if there are subject matters that are impossible to completely formalize, they are impossible to completely formalize in computer programs.

There are unformalizable forms of expertise, such as hands-on skills. Even something as apparently rigid as playing chess turns out to be based on much tacit knowledge. Chess players at the highest level do not apply theories of the game, and they do not even examine more possible moves than players directly below them in skill. Chess players become better by recognizing situations as relevantly the same as ones they have seen before, and knowing what to think about for those situations. Chess programs have become very good, but they operate on very different principles, applying programmed strategic rules of thumb to enormous numbers of possible moves. For Dreyfus, tacit knowledge is embodied knowledge: the good chess player simply perceives patterns and similarities, and that ability has become part of his or her perceptual apparatus. Computers need to have definitions of patterns before they can be useful, and those are difficult to find, for chess and for most other domains of interest.

H. M. Collins also has a critique of AI (e.g. Collins 1990), and it is not surprising that tacit knowledge plays a role. Collins's earlier work on replication (Collins 1991 [1985]) showed that expertise even about straightforward things is difficult to formalize. This leads him to ask a different question than does Dreyfus: how does AI ever work? His answer is that AI works precisely in those areas where people have chosen to discipline themselves to work in machine-like ways. Even then, computers give the appearance of working well because people easily, and sometimes unconsciously, correct their errors. Pocket calculators work well because people have forced arithmetic into rigid and formal modes. Yet it is not unusual for calculators to give answers like "6.9999996" in response to problems like "7 ÷ 11 x 11"; such mistakes are quickly corrected. More common are "mistakes" like the following. To know your weight in kilograms if it is 165 pounds you type in "165 x 0.454" – and are given the answer "74.91." In normal circumstances, though, the "correct" answer is "75" kilos, especially given that the starting point, 165 pounds, is marked as an approximation by its being a multiple of five.

The developed version of this analysis, in work with Martin Kusch, starts from actions, not knowledge (Collins and Kusch 1998). "Polimorphic" actions – the spelling is right, drawing attention both to multiplicity and to their social derivation – can be instantiated by a number of possible behaviors, only some of which are appropriate in any specific context. Writing a love letter, voting with one's conscience, and being playful are all polimorphic because the actions cannot be completed merely by copying previous instances of the behaviors that instantiate them: a love letter that is merely a copy of a previous love letter would be a failure or a deception; a joke told over and over again ceases to be a joke; a vote with one's conscience has to be related to reasons. "Mimeomorphic" actions, on the other hand, are actions that can be performed by copying some limited repertoire of behaviors. While a computer-generated love letter does not, in general, count as a love letter, a computer-generated invoice counts all too well as an invoice – in fact, computer-generated invoices are coming to look more appropriate than handwritten ones.

Whereas Dreyfus argues that domains in which we have no theory are difficult to formalize, Collins and Kusch argue that socially situated domains, in which human participants have not forced themselves to behave like machines, are impossible to formalize. These two claims are not, strictly speaking, contradictory, but they point in very different directions.

Experimenters' regress

Experiments are normally thought of as offering decisive evidence for or against hypotheses. Because they are supposed to be repeatable, experiments look as though they provide something like solid foundations for scientific knowledge. Collins's notion of *experimenters' regress* challenges the distinctiveness of experimentation. The idea has been widely discussed, and has shaped many arguments in S&TS. At the most abstract level it is another application of the Duhem–Quine thesis, but it is a particularly well-studied application.

At genuinely novel research fronts experimenters do not know what their results are going to be; that is, after all, one of the purposes of experimentation (Rheinberger 1997). Thought of in traditional terms, experimental systems are tools for producing differential responses: we can think of typical experimental systems as devices that reflect the different natures of different inputs, and in so doing have something to say. Medical researchers might have 50 patients take a drug thought to help prevent heart disease, and 50 patients take a placebo. If the first group has significantly less heart disease than the second, then the drug can be said to have the property of preventing heart disease, and thus the drug and the placebo are in this respect different. The experimental system – the patients taking pills in relatively controlled circumstances – is supposed to answer a specific set of questions about the drug. There are many other dimensions to experimental work, but the production of differential responses to different inputs is where the epistemic weight of experimentation is thought to lie.

Experimental systems created for doing real research (as opposed to teaching or demonstration) are generally relatively novel, because they are created to answer questions that have not yet been answered. Based on what we have seen so far, we could expect that it is often difficult to get an experimental system to work, to fine-tune it so that it responds in the right way. Because the TEA-laser was a piece of technology supposed to serve a particular purpose, at least it was relatively straightforward for the researchers to know when the laser was working: it burned through concrete. But how do researchers know when an experimental system is working?

If experiments are supposed to answer genuinely open questions, then the experimenters, and the scientific community in general, cannot know what the answers are, and so they do not have a brute-force way of telling when the experimental system is working. One knows that the experimental system is working when it gives the right answer, and one knows what the right answer is only after becoming confident in the experimental system. Collins calls this problem the problem of experimenters' regress.

Experimenters' regress becomes particularly vivid when there is a dispute. In the early 1970s Collins studied attempts by physicists to measure gravitational waves. The theory of general relativity predicts that there should be gravitational waves, but the general agreement among physicists at the time was that they would be too small to measure using any equipment available at the time. One researcher, however, Joseph Weber of the University of Maryland, developed a large antenna to catch gravitational waves, and started finding some that were many times larger than expected. This started a small flurry of attempts to replicate or disprove Weber's results. Some came out in his favor, and some came out against him. Who was right?

From Weber's point of view, people had attempted to replicate his results too quickly. Whereas he had spent years calibrating his antenna – one experimenter who had worked with him said that Weber "spends hours and hours of time per day per week per month, living with the apparatus" (quoted in Collins 1991: 91). People who could not detect anything with their quickly developed gravitational wave detectors simply published their results, apparently refuting Weber. But were their detectors working well? On the other hand, Weber's results might have been artifacts of a device that was not measuring what he thought, or that was simply behaving erratically. His hours spent making his detector produce signals would be irrelevant if there were no waves to detect.

When should one experiment count as a replication of another? Each of the gravitational wave detectors in this controversy was different from each of the others. This is not merely because it is essentially impossible to create complex novel devices that can be considered exact copies of each other. Scientists rarely want to copy somebody else's work as exactly as possible. Instead, even in the uncommon instances when they are trying to replicate somebody else's experiment, novelty is a goal: they want to refine the tools, try different tools and different arrangements, and apply particular skills and knowledge that they have. So there are no strict replications, replications from a God's-eye point of view.

As the problem of foundationalism (box 2.2) suggests, it is in principle possible to undercut the support of any claim, with enough work. Experimenters' regress is thus a theoretical problem for all experiments, and only sometimes becomes a real or visible issue. Most of the time experimental results are uncontroversial relative to the amount of work it would take to challenge them effectively. If an experimental system is well established or well constructed, the results it produces will be difficult to dislodge.

Culture and Power

Most early laboratory ethnographies addressed philosophical questions about the production of knowledge. A somewhat different strand had its origins in anthropology, and addressed questions about the cultures of laboratories *per se*. Sharon Traweek's *Beamtimes and Lifetimes* (1988) was the result of lengthy ethnographic studies of facilities in which high-energy physics is done; in particular, she contrasts a US and a Japanese laboratory. Traweek studies the locations of power in these laboratories, and particularly how the physical spaces of the laboratories code for and enable that power. She investigates how the high-energy physics culture tells "male tales" that create stereotypes of physicists as men (see also Keller 1985; Easlea 1986).

Following traditional anthropological interests, Traweek explores physicists' different understandings of space and time, and in particular how different understandings of time – such as the time of physical events, reactor time, and trajectories of careers – interact. For Traweek, then, laboratory ethnographies are investigations of interesting cultural spaces of the modern world.

Diana Forsythe's studies of artificial intelligence laboratories (Forsythe 2001) amplify some of the results of earlier laboratory studies in the context of software engineering. Because of the sites she studies, she can juxtapose more anthropological understandings of knowledge with the AI researchers' understandings of knowledge as mere collections of facts. The value in that juxtaposition stems from what it reveals about the culture in the subject laboratories, and the power relations between AI researchers and the people whose knowledge they are trying to formalize. Her interest in power relations also leads Forsythe to some interesting observations about "studying up" in ethnography, studying people with more status and power than the ethnographic researchers (Forsythe 2001). A number of Forsythe's studies are of attempts to create expert medical systems. It is common among the AI researchers to "blame the user" for the failings of the expert systems, and more prominently for the failure of the medical establishment to use these systems. But the AI researchers' view, a feature of the cultural landscape of AI, of knowledge as something that can be extracted from experts (typically in interviews), results in an insulation of the systems from the real medical contexts in which they are meant to operate. Because AI research has relatively high status, the researchers can see the doctors as informants whose expertise can be mined for their programs. And patients' expertise is ignored altogether.

Most of the first round of laboratory studies focused on local actions of and interactions among actors trying to create knowledge, and emphasized the contingency of local situations. But they largely left aside institutional features that shape what can be knowledge. Daniel Lee Kleinman (1998), studying a university biology laboratory, argues that industry patronage of fields affects laboratory practice at very fine levels. A bacterial agent being studied in the plant pathology lab in which he did his research was evaluated in a particular agricultural context, subjected to tests against standard *chemical* fungicides to see its effectiveness against agriculturally important root rots. Kleinman also argues that standardized technologies, which affect research intimately, are often embodiments of power relations: laboratories may not be in a position not to use them, or to use them in non-standard ways. And he argues that research is shaped by assumptions about patents and about laboratories' relations to the universities in which they sit. The institutional landscape should figure more in laboratory studies.

Extensions

The laboratory need not be neatly bounded by four walls, and laboratory studies need not describe only experimental work. Researchers have therefore extended the approach, problems, and style of early laboratory studies to other areas.

Martina Merz and Karin Knorr Cetina (1997; see also Gale and Pinnick 1997), for example, take laboratory studies to theoretical physics, looking at some collaborative work at CERN, the European Laboratory for Particle Physics. Physicists employ a variety of heuristics and tricks for expanding their theoretical objects into combinations of simpler ones. None of these takes the form of rule-bound procedures, which makes experience and intuition about *how* to apply the heuristics valuable. Like the problems that experimenters face, the problems that Merz and Knorr Cetina's subjects face cannot be solved by following neat recipes, but only by exploration, tinkering, sharing expertise, and eventually hitting upon the right answer.

The results of laboratory studies also translate to field contexts. In a neat study of agricultural field trials, Christopher Henke (2000) shows how scientists are dependent upon the skills and cooperation of farm workers and farm owners; they therefore have to devote considerable effort to assuring that cooperation. Latour's examination of some ecological research in Brazil argues that a substantial part of the work of field research is work to turn the field into a laboratory (Latour 1999). Correspondences between the natural world and theoretical concepts can only be drawn if the natural world is physically transformed into representations. In a very different vein, but producing a similar result, Charles Goodwin's (1995) study of an oceanographic research vessel investigates how researchers from different disciplines work together to obtain samples. Like Traweek, Goodwin is concerned with representations of space, and with the coordination of multiple spaces. Correspondences have to be drawn between the bottom of the ocean floor, spaces created by the sonar that detects features of that ocean floor, spaces of the computer screen and spaces established by theoretical work. These correspondences require physical probes and markers that can be identified in different representations; they demand complex coordination by the researchers.

Laboratory studies have even been taken to the social sciences. Some of the work of Malcolm Ashmore, Michael Mulkay, and Trevor Pinch's (1989) study of health economics depends heavily on laboratory studies. Daniel Breslau and Yuval Yonay's (1999) study of economic modeling shows that it is quite analogous to experimental work, involving skill and tacit knowledge, give-and-take with the models, and shaped by important cultural or technical assumptions. Rob Evans (1997) shows how macroeconomic models are flexibly interpretable, and are difficult to falsify. Douglas Maynard and Nora Cate Shaeffer (2000) take S&TS to sociology, observing telephone surveys. Again, skill plays a key role, even when the survey is ideally a purely formal instrument: good questioners deviate from their scripts in order to make their interactions with respondents more closely follow the formal model. Even the apparently unlikely site of science policy has been approached via ethnographic methods, showing how science policy is constituted by a set of representational practices (Cambrosio, Limoges, and Pronovost 1990).

The laboratory study, then, is moving into new terrains. There, original results are often seen again. However, especially with an interest in specific cultures, and in power, every new terrain is also a productive site for new insights.

CHAPTER 10

Controversies

Opening Black Boxes Symmetrically

Science and technology appear to accumulate knowledge, piling fact upon fact in a progressive way. Although this may not be an entirely accurate picture – Kuhn argued that science is less progressive than we take it to be, that there are losses in knowledge as well as gains – scientific and techno-logical discussions seem to end with claims to solid knowledge more often than do discussions in most other domains. In this context science and technology studies (S&TS) appropriates the engineers' term *black box*, a term describing a predictable input-output device, something the inner workings of which need not be known for it to be used. Science and tech-nology produce black boxes, or facts and artifacts that are taken for granted; in particular, the history and grounds of their becoming good facts or suc-cessful artifacts is seen as unimportant to their use.

Once a fact or artifact has become black-boxed, it acquires an air of inevi-tability. It looks as though it is the only possible solution to its set of prob-lems. However, this tends to obscure its history behind a teleological story. If we want to ask how the consensus developed that vitamin C has no power to cure cancer (see Richards 1991), it hinders our investigation to say that that consensus developed because vitamin C has no power to cure cancer. The truth has no causal power that draws scientific beliefs toward it. Instead, consensus develops out of persuasive arguments, social pressures, and the like. Similarly, if we want to ask how the bicycle developed as it did, it hin-ders our investigation to claim that bicycles with equal-sized wheels, pneu-matic tires, particular geometries, etc., are uniquely efficient. The implausibility of that claim can be seen in the wide variety of meanings attached to bicy-cling and consequently the wide variety of shapes of bicycles over the past 150 years (see chapter 8; see also Pinch and Bijker 1987; Rosen 1993). Bicy-cles and bicycling are *interpretively flexible*, having different meanings for different actors. There is no standard of efficiency that draws artifacts.

To disperse the air of inevitability, S&TS takes a symmetrical approach to studying debates. Rather than look at facts and artifacts after they have been black-boxed, investigators pay particular attention to controversial stages in their histories. This might even involve studying genuinely open controversies, controversies that have yet to end. Thus researchers in S&TS have looked at water fluoridation (Brian Martin 1991), parapsychology (Collins and Pinch 1982), the development of electric cars (Callon 1987), the best strategies for artificial intelligence (Guice 1998), the definition of the moment of death (Brante and Hallberg 1991), and many other recent and current controversies. The model has even been extended to cover ethical controversies, such as that over embryo research (Mulkay 1994) or over genetic sampling of human populations (Reardon 2001).

Reasonable Disagreements

When looking at controversies, there is a temptation to see the losing participants as unreasonable. However, we might take a more charitable attitude especially toward robust disputes over scientific and technological issues. The participants in disputes always have reasons for their positions, and they, at least, see those reasons as good ones. A symmetrical approach attempts to show some of the force of those reasons, even the ones that eventually fail. Thus S&TS attempts to recover rationality in controversies, where the rationality of one or another side is apt to be dismissed or forgotten.

For example, in the 1970s the geophysicist Thomas Gold proposed an "abiogenic" theory of the formation of hydrocarbons. Oil, he argued, is not a fossil fuel, but is a purely physical product. From Simon Cole's (1996) essay on the controversy that ensued, it can be seen that Gold's model of the formation of oil can be defended, and that serious questions can be raised about petroleum geologists' standard models. While Gold may be arguing for an unorthodox position, and perhaps even one that is difficult to maintain, it is not an irrational one.

Once again the Duhem–Quine thesis is relevant. That thesis states that it is always in principle possible to hold a theoretical position in the face of apparently contrary evidence (see box 1.2). Though this looks like an epistemologist's claim with little applicability in practice, it is relatively easy to document cases in which scientists and engineers reasonably maintain positions – by plausible standards of reasonableness –   despite evidence that other scientists and engineers find conclusive against those positions. The same can even be seen in mathematics, often thought to be a purely formal activity with no possibility for reasonable disagreement (Crowe 1988; MacKenzie 1999; Restivo 1990). Imre Lakatos's classic *Proofs and Refuta-*

tions (1976), for example, shows how the history of one mathematical conjecture is a history of legitimate disagreement (see box 5.1 for Bloor's sociologizing of Lakatos's case).

At the same time, most minority views are eventually excluded from public debates. Extremely deviant views are marginalized, and more moderate deviants may be appeased and answered. (Though discussion of controversies is often framed in terms of neatly opposing sides, there are not usually just two sides in a controversy – Jasanoff 1996.) Disagreements are regularly and routinely managed and contained. Though challenges may be unanswerable in principle, these challenges are in fact answered, assuaged, or avoided, allowing agreement and knowledge to be built up.

Box 10.1   Cold fusion

In 1989, Stanley Pons and Martin Fleischmann, two chemists at the University of Utah, announced at a press conference that they had observed nuclear fusion in a table-top device. The announcement was stunning. It had been assumed that fusion, the process that produces the sun's energy and that is at the heart of a hydrogen bomb, could only take place in situations of very high energy, and could only be controlled with large and elaborate devices. Huge sums had been invested in conventional fusion research, largely in the hopes that some method of producing controlled fusion reactions would lead to almost limitless energy. Pons and Fleischmann were apparently offering an inexpensive alternate route to that energy.

The fusion apparatus was simple and apparently easy to create. Many other groups rushed to repeat the experiments, initially working only from television and newspaper descriptions and images. Some of these groups immediately confirmed the initial results; negative results appeared more slowly, but the tide turned against cold fusion in the months that followed. Meanwhile, theoretical arguments were created to show how cold fusion could have occurred, while others were marshaled for the impossibility of cold fusion; again, the tide turned against cold fusion. Disciplinary divisions were immediately relevant, because fusion had been the preserve of physicists until that point, and many of them were quick to dismiss the two chemists who were cold fusion's primary discoverers. In a matter of months the skeptics had prevailed, and consensus was rapidly forming that cold fusion was the erroneous product of incompetent experimenters.

A number of researchers in S&TS have looked at the debate that followed Pons and Fleischmann's announcement, from a variety of perspectives. Bruce Lewenstein (1995) uses the cold fusion case to trace how information and views of cold fusion circulated among researchers. The standard model of scientific communication emphasizes formal publications, but during this controversy researchers relied on every medium available to learn what was happening, including electronic bulletin boards, television and newspaper reports, and faxed preprints. The frenzy of communication helped to solidify views, as people learned about trends in the theoretical and experimental arguments. Harry Collins and Trevor Pinch (1993) and Thomas Gieryn (1992) use the controversy to illustrate the difficulties of replication of experiments and to illustrate how the controversy was resolved without iron-clad proofs or refutations. Perhaps most interestingly, they illustrate the interpretive flexibility of the experimental work. Physicist Steven Jones at Utah's Brigham Young University, who had been in close contact with Pons and Fleischmann, and had been working on similar experiments, presented his work much more modestly. Collins and Pinch argue that had Jones prevailed in the interpretation of the results, cold fusion might have survived under a very different label, as an unexplained phenomenon worth exploring, though not one of earth-shaking importance. Bart Simon (1999) argues that while consensus did quickly form against cold fusion, that consensus did not stop research. The controversy, and as a result cold fusion science, was dead within a year, but many otherwise reputable researchers kept on doing experiments – hidden from mainstream science, being performed after normal working hours or in garages. Simon thus draws attention to the distinctness of official science's consensus and the views that people hold.

Interests and Rhetoric

Controversy studies make manifest the processes that lead to scientific knowledge and technological artifacts. In the midst of a controversy, participants often make claims about the stakes, strategies, weaknesses, and resources of their opponents. Therefore, researchers in S&TS have access to a wider array of information when they look at periods of the controversy than when they look at periods after controversies have been resolved.

What leads the central protagonists of a controversy to take their positions, particularly when they are unorthodox positions? This question is only sometimes asked explicitly. Interestingly, not asking it amounts to assuming that the positions adopted by key participants are of intrinsic value, that those participants adopt the perspectives they do simply because they find them attractive or plausible. Why does Thomas Gold take on well-entrenched beliefs and claim that petroleum is the result of purely geophysical processes, not biological and geophysical processes combined? Because of the value of his hypothesis? Because he enjoys challenging orthodoxy?

If we ask the question explicitly, we may turn instead to either interest models or contextual explanations. For example, intellectual positions may cohere with social positions. In Donald MacKenzie's (1978) study of a dispute over statistical tools, a case is made that one of the competing tools better served the interests of its originator Karl Pearson, because it served the interests of the professional middle class, being of particular interest to eugenic projects. Positions may also cohere with past investments in skills, resources, and claims: positions fit better or worse into programs of research. In Henrika Kuklick's (1991) analysis of a dispute over the origins, dates, and importance of Great Zimbabwe, an archaeological site, various professional interests and established positions clash with each other, mixed with familiar conflicts over land and race. Looking at technological controversies, the interest model can become much more straightforward. In a biotechnology patent dispute studied by Alberto Cambrosio, Peter Keating, and Michael MacKenzie (1990), the sides are competing companies each of which has a strong financial stake, stemming from investments in research and expertise, in the outcome.

What are the tools actors employ to further their positions? Scientists and engineers need to convince the appropriate people of their claims, and therefore the central tools on which to focus are rhetorical ones. In science in particular, some of the rhetoric is easily available for study, because a main line of communication is the published paper, an attempt to convince a particular audience of some fact or facts. In both science and technology, less formal lines of communication such as face-to-face interactions are less amenable to study, but they are no less rhetorical; thus when they can be studied they can also be studied in terms of their persuasive power (see chapter 14).

This emphasis on persuasion is not to deny the importance of the creation of new pieces of evidence, for example performing new experiments to support one or another claim. However, viewed from the perspective of published papers, rhetoric always mediates material actions like experiments and observations, standing between readers and the material world. That is, experiments and observations are presented to readers, so they are inputs to the rhetorical task. In addition, empirical studies are designed with persuasion in mind: a study that has no potential to convince audiences is a poor one. Since rhetoric is the main subject of a later chapter, the discussion here will be brief.

A central rhetorical task in a controversy is to convince audiences of the legitimacy of one's own positions, and the illegitimacy of those of opponents. To a large extent this means making one's own work more scientific, or more central to key traditions, than that of one's opponents. Thus one of the most important rhetorical resources is the idea of science itself. H. E. Le Grand (1986) describes how in the dispute over plate tectonics a number of the disputants invoked maxims about the scientific method to appropriate the mantle of science. Some participants portrayed themselves as more concerned with fidelity to data and thus as more empiricist; some portrayed themselves as making their claims more precisely falsifiable; and some took the risky strategy of allying themselves with a Kuhnian picture of science.

Similarly, disciplines are important. In the cold fusion saga, for example, the fact that two of the main proponents of cold fusion were chemists made them easy targets for their physicist challengers (see box 10.1). The physicists could point to the tradition of physical interest in fusion, and dismiss the radical statements of a few outsiders. Meanwhile, the chemists could point to physicists' interests in the large sums of money available for conventional fusion research. More subtly, authors can invoke traditions simply by citing important canonical figures and portraying their work as following established patterns.

Appeals to reputational issues can serve something of the same function. Someone with a reputation for being a brilliant and insightful theorist, a careful and meticulous experimenter, or who simply has a record of accepted results, might be believed on those grounds. Having worked with a respected colleague, being at a large research institution, and having a large laboratory are also aspects of reputation that might be invoked in the course of a controversy, as might opposite claims.

A very different way that authors can legitimate and delegitimate is by invoking norms of scientific behavior. To question other researchers' open-mindedness, whether by showing their commitment to broad programs of research or in the extreme by suggesting that they have financial stakes in the outcomes of research, is to question their disinterestedness. Even humor and creating the appearance of "farce" can be used to isolate positions (Picart 1994). Norms of scientific behavior, such as the ones Merton identified, are resources that can be mobilized in the service of particular ends.

More broadly, however, a scientific paper is designed to convince its audience. In a well-constructed article the data it brings to bear, the other articles cited, its lines of argument, the language it uses, and so on are all designed to convince readers, though they will also reflect the cultures in which they are written. Scientific writing, like most other writing, is constructed to have effects, and when it is carefully done all of its elements contribute to those effects. For example, an article may contain criticisms of assumptions, studies, experiments, or arguments made by opponents in the controversy. This criticism may be blunt or subtle, straightforward or technical – Latour (1987) points out that as issues become more controversial they tend to become more technical, but even technical criticisms are constructed to convince people.

Technological Controversies

More purely technological controversies follow similar patterns. Whether at issue is nuclear energy (Jasper 1992), the choice between different missile guidance systems (MacKenzie 1990), or between electric and gas-powered automobiles (Kirsch 2000; see box 8.2), technical controversies do not have straightforward technical solutions, at least not if those solutions exclude human factors. For example, Cambrosio, Keating, and MacKenzie (1990) show how technical issues in a patent dispute are simultaneously social, historical, economic, and philosophical issues. As in many other patent disputes, the issues revolved around the question of whether a commercially valuable innovation was "novel" and "non-obvious" – whether it should count as a genuine invention or merely the use of an old invention. There is no merely technical answer to this question, in the sense of one that stands apart from issues about the nature and constitution of the expert community, from issues about standards of novelty, and from issues about the histories of their development. There is no neat way to cut through the complexity to be left with a simple, definitive, and non-human solution.

The result of this complexity is that when people, whether they are engineers, investors, or consumers, face a choice between technologies, there is no context-independent answer as to which is better. The choice between a Macintosh and a PC, to take an all-too-familiar example, is a choice between two constellations of interests, goals, claims, images, and existing and potential machines. The choice becomes even more complex when even the most narrowly technical characteristics of competitors may be in dispute: tests to determine those characteristics are never definitive in and of themselves (see box 5.2). And because technological issues tend to impinge on public concerns much more directly than do scientific issues, external actors can quickly become involved in technological controversies. Thus what particular technologies' capabilities and characteristics are can become an issue to be resolved by non-experts as well as experts. Whether non-lethal weapons, such as pepper spray or rubber bullets, are safe or dangerous becomes an issue decided – or left ambiguous – by the interventions of a wide variety of interested parties (Rappert 2001).

Box 10.2   The *Challenger* launch decision

Diane Vaughan's prize-winning *"Challenger" Launch Decision* (Vaughan 1996), is a thorough and incisive examination of a single technological decision, albeit one with large consequences. The January 28, 1986, explosion of the Space Shuttle *Challenger* is a well-known event, in part because of the vivid image of its explosion, in part because of the inquiry that followed, and in part because it seemed to symbolize the vulnerability of the most sophisticated technological systems to small problems. The immediate cause of the accident was the failure of an O-ring that was supposed to provide crucial sealing between the joints of the solid rocket motor – a point dramatized by physicist Richard Feynman in the subsequent inquiry (Gieryn and Figert 1990). NASA had asked the manufacturer of the solid rocket motor, engineering firm Morton Thiokol-Wasatch, about the effects of cold weather during the launch, and the Thiokol engineers responded by pointing out the possible problem. The interpretation supporting their concern was questioned, however, and the launch eventually went ahead.

Vaughan's study is a careful attempt to understand the culture at NASA, and how the decision to launch the *Challenger* was made, given the safety concerns. Ultimately, Vaughan argues that the decision was fully in accordance with NASA protocols, in conjunction with the data on O-ring failure, and a normalization of risk following previous O-ring failures. Blame lies in the culture and organizational structures of NASA, which too-explicitly recognized the risk involved in every decision, and which were not prepared for the problems they faced in January 1986.

William Lynch and Ronald Kline (2000) make interesting use of Vaughan's conclusions in the context of engineering ethics. They argue that simply introducing engineers to moral vocabularies and moral reasoning is unhelpful, because that approach typically leaves dissenting engineers with only an unattractive role to play, that of the whistle-blower. Attention to ways in which procedures can make every step of an unacceptable path rational, and to ways in which risk is incrementally normalized, may be more valuable than attention to grand moral confrontations.

Vaughan's position is not itself uncontroversial, a fact that may demonstrate the resilience of controversies. Edward Tufte (1997), in an argument on the importance of good visual displays of information, argues that had the engineers studying the O-rings created a better graph of the data on O-ring failure, it would have become clear that conditions of the launch were very dangerous ones, and the Shuttle would not have been launched. Many people find Tufte's new graphs striking enough to cut through Vaughan's complex arguments. At the same time, Tufte's graph has itself been criticized for subtly changing the information, and the result is that it is unclear whether such simple blame can be assigned. So the meta-controversy remains unresolved.

The Solution of Controversies

Given the Duhem–Quine thesis, and the impossibility of unchallengeable foundations, how are disputes in science and technology resolved? How is *closure*, the ending of active debate, achieved? Here are five different types of contributions, most of which will be brought into play in any given dispute.

1 Detailed critiques of observations, experiments, and positions

The most straightforward attempts to discriminate among positions start from criticisms and questions about such things as the consistency and plausibility of positions, the solidity of experimental systems, and the appropriateness of experimental or observational procedures. To be taken seriously in the first place, an experiment must be designed so that there are few obvious unanswered questions that could affect how results are interpreted. Nonetheless, as the underdetermination thesis teaches us, it is never possible to answer all questions. Determined opponents can place any part of the system under the spotlight and challenge its role: could it contribute something to the results that undermines their fidelity or trustworthiness? Of course challenges can be answered, too. As one of Collins's subjects in the gravitational wave study (see chapter 9) said, "In the end . . . you'll find that I can't pick it apart as carefully as I'd like" (Collins 1991: 88). The force of a challenge will depend upon the relative balance of its and the response's plausibility, given the work that participants are able to do.

2 Calibration of instruments and procedures

Even without specific criticisms, calibration can help decide among results. Especially when they are complex or delicate, the accuracy and precision of instruments can be challenged. They might be measuring something different from what they are designed to measure. Or, they might simply need adjustment. To demonstrate the (un)trustworthiness of an instrument, one can test it on known quantities. However, just as there can be no perfect replications of an experiment, there can be no perfect calibrations of an instrument: the circumstances in which it is tested and refined are never exactly the same as the circumstances in which it is used (see box 5.2).

3 Isolating one position as more scientific or central – or as deviant

The actions and strategies above are designed to legitimate and delegitimate different pieces of research. We can see that as a more general category for resolving scientific and technical conflicts. Thus the strategies to make work more or less scientific that were described in the last section can also help end discussions. It is particularly important to solidify opinion in the *core-set*, the densely connected group of researchers whose opinions count most (Collins 1991). Public consensus in the core-set can officially end a controversy even while less central researchers maintain deviant views. If the networks are strong enough that everybody who matters understands the consensus, deviants may even continue publishing their views, but find themselves entirely ignored (Collins 1999).

4   Showing one position to be more useful

Separate from issues of validity and legitimacy are issues of pragmatics. Sometimes one idea will become dominant because many researchers can see how to use it, how to build on it, regardless of its validity. Pickering's (1984) study of the triumph of Gauge Theory in physics, the dominant model of particles, places weight on the fact that many physicists already had the mathematical skills and knowledge needed to investigate Gauge Theory, and therefore found it congenial.

5   Ignoring deviant viewpoints and data

Finally, a position that contradicts or runs against the grain of established scientific understandings is less likely to be accepted than one that agrees with expectations. If a position contradicts quite central beliefs then its proponents usually have to do a considerable amount of work for anybody to treat it seriously enough to bother arguing about it. In some cases disputes about deviant ideas or results never arise, or are quickly forgotten, while most people in the field go about their own business. The potential dispute does not become an issue for the community as a whole, because rather than being refuted some results are simply ignored.

How to Understand Controversy Studies

Because S&TS looks at science and technology from the outside, applying frames of analysis not indigenous to science and technology, work in S&TS is sometimes viewed with suspicion. In particular, some scientists and engineers who read work in S&TS think that the field or some of its concepts are attacks on them. Experimenters' regress (chapter 9) is a case in point. Allan Franklin, a physicist and philosopher, reads Collins's work on experimenters' regress as suggesting that experimental results are not to be trusted, and that the resolution of conflicting experimental results does not involve rational discussions (Franklin 1997). Looking at published reports, he reanalyzes the gravitational wave case, and argues that the evidence strongly supported Weber's opponents. In this he is undoubtedly right: most of the scientists involved took the matter as firmly settled, which means that they took the evidence as strongly against high fluxes in gravitational waves. If we look at the episode from the point of view of a participating scientist trying to figure out what to believe, we should expect to find ourselves evaluating much of the same evidence that was in fact used to settle the debate. The point of most work in S&TS, though, is not to simply check or challenge scientific and technical knowledge, but to understand its sources and meanings.

Readings like Franklin's are understandable, however. Symmetrical presentations of controversies are intended to show that disagreements are legitimate, that there is a case for the heterodox positions. Because they are often intended as arguments against rigid views of the scientific method, controversy studies are intended to show that there is no decisive evidence for or against scientific claims, that there is no agreed-upon formula that inexorably leads to one answer over the other. In addition, the results of controversy studies are not necessarily flattering to science and engineering, when, for example, they point to the unruly processes of arguing.

Nonetheless, researchers in S&TS are typically far too well versed in the details of their cases to suggest that closure of debates does not involve careful reasoning. They do not necessarily want to paint unflattering portraits, and they certainly do not suggest that one should not believe scientific results or use technologies. Controversy studies, at least within standard S&TS frameworks, are aids to understanding the nature of the closure of debates. That is, symmetrical presentations of controversies highlight the solution of debates as local and practical achievements that need to be understood in terms of the local culture. Why did a debate end which could, in principle, have continued indefinitely? If there were decisive interventions, what made them decisive?

When controversy studies work well, then, they show how evidence is tied to its local culture and contexts. By itself, some piece of data has no meaning. Data are only given meaning – as evidence – by the people who make use of them. Studies of scientific controversies show how people can give meaning to information and how they sometimes convince members of a community to agree with that meaning. They show how knowledge is built by a process of non-foundational bootstrapping, but not how knowledge is groundless (Nickles 1992).

Captives of Controversies: the Politics of S&TS

While controversy studies may be symmetrical, they are rarely neutral. By showing the social mechanisms of closure, controversy studies tend to be viewed as supporting the less orthodox positions. Therefore, the results of the studies can themselves become part of the controversy, picked up by one or more sides, though probably by the underdogs. Thus controversy studies, and the researchers who perform them, run the risk of being "captured" by participants (Scott, Richards, and Martin 1990). Especially since uses of the study are somewhat unpredictable, this can be worrisome for S&TS researchers.

How likely is it that controversy studies will be appropriated by one or another side in the controversy? Evelleen Richards (1996) reports that her earlier study of the controversy over vitamin C and cancer is viewed as supporting alternative medicine in its struggle to have the positive effects of vitamin C recognized. Organizations that support alternative medicine sell copies of some of her articles, and articles on alternative medicine cite her as exposing "the corruption at the heart of the cancer hierarchy in America," among other things (Richards 1993: 344). This is despite the fact that she sees herself as articulating a position distinct from those of both alternative medicine and the medical establishment. A study of a variety of roughly "New Age" movements by David Hess (1997) was similarly picked up by those movements, and criticized by their opponents.

Harry Collins and Trevor Pinch's (1982) study of parapsychology provoked the same reaction – but Collins also reports surprises in the reactions to his gravitational waves study (Collins 1996). This points to some unpredictability in the "capture" phenomenon, an unpredictability that could just as easily give comfort as discomfort to the student of controversies: in general, published work is available for use, and authors have little control over that use. The most unpredictable aspect of "capture" may be capture itself, because it is unclear what percentage of S&TS controversy studies have affected their controversies. And there are many different levels of "capture," and some may be more or less irrelevant; it may not really matter within a scientific controversy that one or another side makes use of an S&TS analysis.

Nonetheless, the recurring possibility of "capture" makes the non-neutrality of S&TS's work clear. Pam Scott, Evelleen Richards, and Brian Martin (1990) argue that S&TS should not pretend neutrality, and that researchers should make their commitments explicit. This would both allow their work to have more practical value, taking it out of the ivory tower, and allow the researchers to have more control over the uses to which their work is put. Since among the many roots of the field are some activist ones, this recommendation is attractive to some people in S&TS. Many within the field feel that it has largely abandoned its activist roots, and that it should re-establish a commitment to the democratization of science and technology (e.g. Fuller 2000a).

There are different ways of making commitments explicit. The strongest version involves becoming an active participant in the controversy. Brian Martin (1996) reports on an "experiment" in which he helped to publicize an unorthodox theory on the origins of AIDS, the theory that it was transmitted via impure polio vaccines. His participation allowed Martin access to documents to which he would not otherwise have had access, although it also closed off access to some people on the other side of the controversy. Although Martin's experiment looks like an extreme case, it may not be: becoming a participant may be particularly easy when policy concerns are visibly intertwined with technical issues, as in many environmental controversies, medical controversies, and controversies over technologies in the public eye.

A more modest form of commitment involves simply making one's position transparent. Donna Haraway's (1988) metaphor of "situated knowledges" may be helpful here. Haraway argues that it is possible simultaneously to strive for objectivity and to recognize one's concrete place in the world. This "embodied objectivity" can produce partial knowledge, though knowledge that is in every way responsive to the real world. In the context of addressing the problem of "capture," David Hess (1997) argues that accepting Haraway's analysis means that researchers can identify the perspective from which they write and act, without relinquishing their status as objective researchers.

Such a perspective may not be more than a position from within the discipline. The perspective that a researcher in S&TS brings to bear on a controversy is unlikely to mirror that of any of the participants – even a symmetrical approach to the social constitution of scientific and technological knowledge should be distinct from the approaches of participants. One way for researchers to articulate a perspective from which they research and write is to articulate a properly analytical perspective. While that may not be neutral between parties, it should be non-aligned. Arguably, an analytic perspective from within S&TS is the only form that commitment need take, and a full and sophisticated analytical perspective can be a basis for sophisticated political positions.

CHAPTER 11

Standardization and Objectivity

Solidarity

In an influential study of British X-ray crystallography, John Law (1973) connected the social structure of the researchers to the nature of the problems being addressed. X-ray crystallographers aim X-rays at crystals, so that they can observe patterns of diffraction and make inferences about the structure of the molecules making up those crystals. The bulk of early work (to the 1930s) in X-ray crystallography investigated the structure of small organic and inorganic molecules, and the crystalline structure of metals. In the mid-1930s, W. T. Astbury and J. D. Bernal started studying proteins, much larger and less tractable molecules. Because proteins were difficult objects of study, most crystallographers were uninterested in them and the problems that they posed. As a result, Astbury and Bernal and their students became the leaders of a new specialty, X-ray protein crystallography, which had only weak links to the original. In fact, relations between the groups were somewhat antagonistic. One protein researcher recalls: "Many of the professional crystallographers . . . regarded it as a complete waste of time. If any of the protein crystallographers made a mistake, then there was much jeering" (quoted in Law 1973: 297).

Using a distinction from Émile Durkheim, Law argues that uniting the original crystallographers was the fact that everybody was doing similar work with similar tools: they were bound by *mechanical solidarity*. Each of the researchers could recognize and evaluate the quality of the work of each of the others. Law argues that this created a strict attitude toward deviance: problems, and variations on techniques, were either permissible or impermissible. The protein crystallographers, on the other hand, devoted enormous energy to each individual protein that they studied, adjusting their techniques and approaches as the problems they faced demanded. They quickly found themselves interacting not with other crystallographers, but with other people studying the proteins that they were interested in. Thus they became part of the "protein community," which was bound together not by techniques but by subject matters. The protein community had *organic solidarity*, in which labor was divided around common ends. Organic communities are more tolerant of deviance: problems and techniques are fit into a gradation of more and less preferred, depending on their likelihood of contribution to common ends.

The framework and terminology that Law uses is not value-neutral. It is difficult not to see mechanical solidarity as a less desirable basis for social interactions than organic solidarity, limiting freedom and creativity. So it is worth balancing Law's study with approaches that show the advantages of mechanical solidarity in scientific specialties, and the pressures toward it.

Getting Research Done

In an analysis of scientific styles, Stephan Fuchs (1992) combines two variables, *task uncertainty* and *mutual dependence*. Because research in the natural sciences tends to be more expensive than that in the humanities and social sciences – tools and materials are more expensive – researchers in the natural sciences tend be more mutually dependent. If tasks are relatively routine, in addition, then there is considerable pressure to produce *facts*, as opposed to interpretations: scientists will be interested in and potentially able to use each other's results, and therefore those results are made easily isolable and transportable. The natural sciences thus often participate in a fact-producing cognitive style, which shapes all aspects of work.

New and innovative research areas are by definition less routine; they therefore have a higher task uncertainty, and according to the theory should probably have a less authoritative style. Disciplines with low material demands or well-distributed resources should have less mutual dependence, and consequently less social pressure to solidify results into factual nuggets; authority can be fragmented, and built on relatively local grounds, since there is less need to share resources. Disciplines with neither well-established tasks nor high levels of mutual dependence should have less stability, and hence more likelihood of continuing disagreement. Fuchs's framework suggests how communities that practice more standardized research are likely to be more coherent and successful. Having discovered how to create feasible research projects, they can both create and pursue them.

A similar analysis can be found in Joan Fujimura's (1988) study of cancer research. In the 1980s the molecular biological approach to fundamental cancer research overtook all of the others (box 11.1). To use Fujimura's term, there was a molecular biological "bandwagon" in cancer research. Fujimura, who starts from the premise that science is work, is interested in how a need to create stable work helped to shape cancer research. She argues that this bandwagon was facilitated by the development of a standardized package consisting of a theoretical model for explaining cancer and a set of technologies for exploring the theory. A standardized technology is one that has become a black box (Latour 1987), or for which the contexts of successful use are flexible (Collins and Kusch 1998). Standardized technologies are shaped such that the skills required to use them are either very generally held or quickly learned.

Scientific research, especially laboratory research, takes funding: equipment, samples and materials have to be bought, and research assistants have to be paid. But funding agencies are much more likely to support scientists who have established track records of successful research. The *cycle of credibility*, in Bruno Latour and Steve Woolgar's (1986 [1979]) terminology, is the cycle that allows scientists to build careers and continue doing research. Continuing research is central to the identity of most scientists, more so than, for example, earning money – even with the rise of a patent culture oriented around commercialization of results, the reward is often funding to do more research (Packer and Webster 1996).

To gain funding scientists typically write grant applications to public and private funding agencies. The success of grant applications depends upon evaluations of the likelihood of concrete results coming out of the research project. That in turn depends upon issues of credibility and direct evaluations of the *doability* of the project itself (Fujimura 1988). Thus scientists have a large stake in finding doable problems, both to gain immediate funding and to build up their base of credibility for the future. A standardized package is attractive because it helps to routinize research, research that is intrinsically uncertain. The pressures of continuing research, then, can easily push toward a type of mechanical solidarity in which researchers in a field do their work in standardized ways. One of the effects of this standardization is that members of a field can see each other's work as meeting standards of objectivity.

Box 11.1   The molecular biological bandwagon in cancer research

According to Joan Fujimura (1988), in the 1980s fundamental cancer research became reorganized around a standardized package of theory and technologies. The theory in question was the oncogene theory, which maintains that the normal genes that contribute to the necessary process of cell division can turn into oncogenes that create cancerous cells dividing out of control. This theory thus offers an explanation of widely different types of cancer, potentially unifying cancer research around a single step in the multiple pathways to cancer. The technologies of the package were recombinant DNA technologies, which have been becoming steadily more powerful since the mid 1970s.

In the early 1980s recombinant DNA technologies become developed to the extent that researchers could be assured of success in sequencing genes. Students could learn to use the technologies relatively easily, and could set tasks that they could be relatively certain of completing within a fixed window of time. For both graduate students and senior researchers this was quite important: graduate students need to know that their research projects will not continue indefinitely, and more senior researchers want to know that they will have results before their grants run out.

Cancer researchers found themselves faced with a standardized package that promised avenues for successful research, avenues for funding, and a possibility of a breakthrough. As more and more cancer researchers adopted molecular biological techniques, and molecular biologists engaged in cancer research, they created a snowball by making one approach the norm, and marginalizing others. The rational researcher joined the bandwagon and contributed to it. It is worth quoting one of Fujimura's interview sources at length, for this researcher summarizes the process neatly:

> How do waves get started and why do they occur? I think the reason for that has to do with the way science is funded. And the way young scientists are rewarded ... [A] youngster graduates and ... gets a job in academia, for instance. As an assistant professor, he's given all the scut work to do in the department. ... In addition, he must apply for a grant. And to apply for a grant and get a grant ... nowadays, you have to first show that you are competent to do the [research], in other words, some preliminary data. So he has to start his project on a shoestring. It has to be something he can do quickly, get data fast, and be able to use that data to support a grant application ... so that he can be advanced and maintain his job. Therefore, he doesn't go to the fundamental problems that are very difficult. ... So he goes to the bandwagon, and takes one little piece of that and adds to a well-plowed field. That means that his science is more superficial than it should be.   (quoted in Fujimura 1988: 275)

Absolute Objectivity

Although "objectivity" is a slippery term, most of the senses of objectivity relevant to the study of science and technology can be grouped in two clusters: *absolute* and *formal* objectivity. In the context of science, absolute objectivity is the ideal of perfect knowledge of some object, knowledge that is true regardless of perspective. Absolute objectivity is the "view from nowhere," in Thomas Nagel's phrase. Formal or "mechanical" objectivity, on the other hand, is the ideal of perfectly formal procedures for performing tasks. In this sense the ideally objective scientific researcher would be machine-like in his or her following of rules.

Given the interpretive nature of scientific knowledge, it is difficult to make sense of the notion of absolute objectivity as anything other than a vague regulatory ideal. Nonetheless, one can trace the history of that ideal. For Stephen Toulmin (1990), for example, the drive for certainty or objectivity has its origins in an attempt to disconnect philosophy and science from political and economic uncertainty. Toulmin takes as an important symbolic – for the young René Descartes – event the assassination of Henry of Navarre, after which there remained little possibility in Europe for political moderation. According to Toulmin, Henry of Navarre stood for a humanist tolerance of divergent views, of which religious views were key to the fate of Europe. In place of the failed humanist tolerance, Descartes attempted to install austerity and distance on the part of the knower, and stances that did not depend in any way on the individual subject.

One interpretation of the ideal of absolute objectivity is in terms of value-neutrality. Robert Proctor (1991) shows that this latter concept has changed with changing historical contexts: it has stood variously for a separation of theory and practice, for the exclusion of ethical concerns from science, and for the disenchantment of nature. In addition, Proctor argues that the value-neutrality needs to be understood in terms of political context. For nineteenth- and twentieth-century German academics, for example, value-neutrality became an ideal in the social sciences because it served to insulate and protect science and the university from external controversies. This was not a neutral stance, but was articulated in opposition to particular social movements and used to combat those social movements. Values should be separate from science because particular values should be separate.

There is no obvious way of making sense of absolute objectivity in practice. However, the judgments that experts make can stand in for absolute objectivity. Expertise is the ability to make (what are perceived as) good judgments, so expertise presupposes a grasping of the truth, if not distanced passionless knowledge. The truth that expertise is supposed to provide is objective truth, containing in itself no contribution from the expert. We will return to this stand-in in the next chapter.

Formal Objectivity

In contrast to absolute objectivity, we can observe approximations of formal objectivity. We can start with an analogue in the world of measures.

Witold Kula's *Measures and Men* (Kula 1986), an historical study of units of measurement and attempts at their standardization, shows the conflict between local and global measures, and the conflict between expertise and objectivity. A good example comes from Bourges in late eighteenth-century France:

> A *seterée* of land is the only measure known in this canton. It is larger or smaller depending on the quality of the soil; it thus signifies the area of land to be sown by one *setier* of seed. A *seterée* of land in a fertile district counts as approximately one *arpent* of one hundred *perches*, each of them consisting of 22 feet; in sandy stretches, or on other poor soils, one *arpent* consists of no less than six *boisselées*.   (in Kula 1986: 30–1)

Before the nineteenth century, many European units of measurement, of different kinds of commodities and objects, were both local in definition and depended upon the qualities of what was being measured. Measurement required both local knowledge, and expertise in how to measure. Movements of standardization were attempts to eliminate that necessary expertise. From a local point of view, though, standardized units were not always preferable. Not only were they unfamiliar, they changed the nature of measurement. For a farmer, the *seterée* above could be a more useful measure than a standardized *arpent*, and certainly more useful than the unfamiliar *hectare*. From a less local perspective, such as that of a national government interested in collecting taxes, the *seterée* is a hindrance to useful, general knowledge.

Something similar takes place in science and technology. Through examples of a few technical units Joseph O'Connell (1993) shows how standardization creates a kind of universality. C. C. Gillispie's *The Edge of Objectivity* (1960) is a sweeping history of the progress of the tools, metrics and frameworks that eliminate subjectivity, and that dictate what is relevant in their objects. There is a sense in which all histories of the progress of scientific knowledge are histories of objectivity, but Gillispie's is unusual in its dramatic yet subtle ambivalence about objectivity: it is a positive virtue, but one that derives in part from the elimination of passion and intimacy with nature.

Even an emphasis on empirical facts participates in the history of objectivity. Lorraine Daston (1991) argues that some European intellectual circles of the seventeenth century attempted to curtail divisive arguing over theoretical speculations by creating a new emphasis on and interest in isolated empirical facts, particularly experimental facts. Although the concept of (formal) scientific objectivity as such is a later invention, community agreements around isolated facts provide surrogates for objectivity, since these facts were not as open to question as were theories. The term "objectivity" does not come to take its current meanings and status until the nineteenth century, though, when expanding communities of inquirers made the elimination of idiosyncrasy more important. Daston and Galison (1992) pick up one part of this story in a study of representational techniques in scientific atlases, particularly of the human body. The search for objective representation was a moral, and not just a technical, issue, and was connected to a new image of the scientist as a paradigm of self-restraint. The choice of specimens for these atlases, necessarily a human choice, became a point of crisis, provoking discussions over the merits of different types and ranges of specimens that should be presented. Interest in eliminating choice also led to increasing use of tools for mechanical representation, such as photography, which could be trusted for their impartiality, if not their accuracy.

Formal objectivity and informal expertise can easily come into conflict: objectivity is a form of regulation that limits the discretion of knowledgeable experts. Whereas much of the recent study of science and technology has pointed to the indispensability of expertise and informal communities, Theodore Porter (1995) is interested in the development of rules and formal procedures that can be widely applied. Particularly in public arenas, objectivity, in the form of ever more precise rules and procedures to be followed, is a response to weakness and distrust. The Corps of Engineers and the United States bureaucracy developed apparently uniform cost-benefit analysis procedures to limit pork barreling and to limit possible disagreement, even though uniformity necessarily introduces its own forms of arbitrariness. In contrast, the Corps des Ponts in technocratic France enjoyed more freedom and could calculate costs and benefits in an atmosphere of relative trust that they were acting in the interest of the public. Objectivity in this sense, as rules for scientists and engineers to follow, can often be seen to result from epistemic challenges. When outsiders mount strong challenges to the authority of scientists and engineers, the response is often to establish formal procedures that unify by removing discretion – Porter's formula is that "democracy promotes objectivity" (Porter 1992a).

To take another example, during the Depression the American Security and Exchange Commission changed the practices of corporate bookkeeping (Porter 1992b). The Commission insisted that assets be valued at their original cost, rather than their replacement costs. To accountants the latter is a better measure of the value of an asset, but is prone to manipulation. The Security and Exchange Commission, though, was interested in maintaining public confidence in accounting practices, not in realistic appraisal. In the right circumstances, then, an apparatus that gives a standardized response to a standardized question can be preferable to a person who gives more nuanced responses to a wider variety of questions, even if there is every reason to believe that standardization gets in the way of truth.

In engineering, standardization of artifacts is a related issue, though the challenges may be more obviously economic and political. A world of interchangeable parts is an efficient world. New parts can be simply bought, rather than commissioned from skilled artisans. Ken Alder traces one line of the history of standardized parts to the efforts of French military engineers of the late eighteenth century to buy standardized weapons (Alder 1998). For example, cannons and cannonballs had to match each other to work, and so had to be of precise standard sizes, even though they were made by different artisans in different places. To enforce precision, the engineers developed the notion of tolerance, upper limits on the deviation from specified shapes and sizes. And to enforce tolerances, they developed ingenious gauges. The result was a set of mechanical judges of performance, which allowed inspectors to judge, and artisans little opportunity to contest, which artifacts were the same as the standards.

Formal objectivity, then, is something that people develop in the right social circumstances. Probably more importantly than in many other domains, scientists' and engineers' efforts to create objective procedures mean that formal tools are often entrusted with considerable responsibilities.

CHAPTER 12

Creating Order, Following Rules

What about Interpretive Flexibility?

The previous chapter explored a formal model of objectivity. When groups of experts face strong challenges they can respond by creating formal rules for their behavior. Even in the absence of challenges, science and technology gain power from the solidarity and efficiency that that objectivity creates. Actor-network theory (ANT) makes a related claim: science and technology gain power from the translation of forces from context to context, translations that can only be achieved by formal rules. Intuitively, this appears indisputable. As was argued in the discussion of ANT (chapter 7), in general the working of scientific and technical theories appears a miracle unless it can be *systematically* traced back to local interactions.

However, we have also seen that in practice formal rules always have to be interpreted. Wittgenstein's problem of rule-following shows that no statement of a rule can determine future actions, that rules do not contain the rules for their own application. This problem is not just a theoretical one: norms can be flexibly interpreted, and so do not constrain actions; mere instructions cannot typically replace the expertise that scientific and technical work demands. Expert judgment is an ineliminable feature of science and technology (and much else).

Thus there are difficulties for the formal model of technical objectivity. The formal rules that apparently make up objective behavior cannot by themselves determine behavior. There is always room for interpretation on the part of the people following rules. And so formal rules do not constrain actions as much as the proponents of objectivity might want to claim; instead, formal rules create new fodder for creative interpretations. In many complex situations formal rules cannot replace expertise with success. There are always cases that, because of crucial ambiguity, instability, or novelty, do not fit neatly into the categories set out by formal rules. Experts may be able to recognize these cases and respond accordingly. Therefore, formal objectivity is not always desirable. This is especially true given that real research operates at the edges of knowledge, rather than merely reiterating knowledge – by definition research is supposed to create novelty, and not merely maintain cultures.

Nonetheless, the humanist model that makes informal expertise central to science and technology (chapter 9) faces an equally substantial problem, precisely the problem that the formal model of objectivity claims to solve. The humanist model allows that expertise can be communicated via social interaction, which makes social interaction a mysterious force. Even if social interaction is the best shortcut for communicating many forms of expertise – which it undoubtedly is – what could possibly pass between people that could not be represented as discrete pieces of knowledge? Questions and misconceptions can be cleared away efficiently, but they are cleared away efficiently because discrete information passes between people. Attention is oriented in the right way, to the most important or most salient issues, but again that is done via information. Any other option appears to invoke a fundamentally mysterious force. However, if science crucially depends upon a mysterious force, then the humanist model will fail to explain the success of science. If science works only because scientists are experts on the natural world, then we need to understand the sources of that expertise and its communication. These sources of expertise are precisely what the humanist model does not offer.

The problem we are left with is how to reconcile objectivity and humanism, how to reconcile positions that insist on the power of formal rules with positions that insist on the importance and ineliminability of expert judgment. Both claim to describe essential features of science and technology, and both of those claims make sense and are backed by evidence. Clearly, we need to be looking at what is added to formal rules to make social activities orderly, what is added to allow people to coordinate actions, what is added to allow for communication.

An Ethnomethodological Solution

One way of attempting to resolve the conflict between objectivity and humanism is through ethnomethodology. As its name suggests, ethnomethodology is an approach to studying methods particular to cultures or contexts. For ethnomethodology, commonsense knowledge is the central object of study, because actors' commonsense knowledge produces social structure. Beyond that, ethnomethodology is a notoriously nuanced approach, one that we will not do justice to here; instead what follows is merely a reading of that approach as offering a solution to our conflict.

Let us return to the problem of rule-following. Wittgenstein showed that rules do not have any quasi-causal power, because any course of action is compatible with a given statement of a rule. But if this is true, why are social activities orderly? How do people coordinate actions? Ethnomethodology approaches these questions by closely examining local interactions, to see how people coordinate themselves, producing and interpreting social order. The interactions are local because if Wittgenstein's argument is right, there are no external forces that can dictate how local actors behave.

At the level of local interactions, people do considerable work to create order. For example, telephone interviews in social scientific survey research are supposed to follow scripts; however, the people being called often do not want to answer the survey, and refuse to answer or evade the interviewer. A skilled interviewer is constantly deviating from the script to bring the conversation back to it (Maynard and Schaeffer 2000). Following the script rigidly stands too large a chance of allowing the subject to not answer the survey, whereas deviations may help convince the subject to keep talking. At issue here is a goal: to get through the survey. This goal might be seen as a rule: achieve the script, in the sense of learning subjects' reliable responses, so that they can be used for objective research. Such a rule, though, does not dictate what the speakers say and do. Instead, its status as a goal makes it a product of what they say and do: completing the script depends on small deviations from the script in order to allow the conversation to continue. People attempt to bring their behavior into line with the rule, but are not governed by it. The result is a future-oriented object, the completed survey, which can be used by researchers to create more knowledge (see Miettinen 1998).

Many rules, then, do not dictate behavior, but instead serve as goals or ideals that skilled actors try to reach. In contrast to the model of objectivity that sees researchers *following* rules, the ethnomethodological model sees researchers variously *achieving* rules. People can accomplish formal behavior, and thus allow for the coordination of actions. Ethnomethodology in this way supplements the model of objectivity, by showing how the formal products that are essential to the objectivity and applicability of science can be created. It also accepts a humanist insight, by accepting that expertise is necessary to achieve formal behavior.

The issue of social order is important because science and technology can be seen as an extremely efficient producer of social order. Every time researchers come to agreement, whether it is about an observation, an interpretation, a phrasing of conclusions, or a theory, they are producing order. That order typically comes to be seen as the result of the order of nature or necessity. It only takes a little bit of study to see that nature and necessity do not routinely force their regularity upon scientists and engineers, and thus the work done to make their orders manifest is a central part of the study of science and technology (e.g. Turnbull 1995).

Creating Orderly Data

How do such insights play themselves out in Science and technology studies (S&TS)? To illustrate them, we might look at the work that researchers do to create order at the level of their data. Here the immediate interactions are not interpersonal, but are with material objects. Nonetheless, the orderly data form the beginnings of work shaped by its potential to be assessed by a larger audience.

"Data" comes from the Latin for "givens," but ethnographic study shows just how much work has to be done to create typical data. While data are givens within science, the creation of data is a topic for investigation in S&TS. In a neurobiology laboratory, for example, photographs of sections are carefully marked to highlight the features that the researchers want to take as data (Lynch 1985). These features are not merely features that would be invisible, indistinct, or unremarkable to an untrained eye, but are typically features that only make sense in particular experimental and observational contexts. That is, the markings and enhancements of the slides generally bring to the fore features that the researchers, working in local research contexts, were looking for. So not only does it take expertise to read such slides – which, following Collins's thinking about expertise, means that there are no mechanical rules for doing such reading – but the expertise needs to be attuned to local circumstances. The careful highlighting and labeling of features on the photographs in some sense takes them out of their most local contexts and makes them available for inspection by the relevant expert community more generally.

While there are generally applicable methodological rules, work has to be done to make them applicable in particular contexts. Even within disciplines, such apparently straightforward things as measurement and observation, let alone more obviously complicated things as maintaining controls in experimentation, interpreting results, and modeling are local achievements (Jordan and Lynch 1992).

Since observation is not a straightforward process (chapter 9), how do data ever become stable enough to form the basis of arguments? How do researchers turn smudges into evidence? To see how, we might look at processes that S&TS has identified, by which researchers in the laboratory decide on the nature of a piece of data, and by which they construct evidence for public consumption.

Many scientific and technical discussions are occasions on which judgments are clarified and organized. Here is a transcription of an exchange between collaborators, taken from Michael Lynch's ethnographic study of a neurobiology laboratory (Lynch 1985: 229):

```
 1 R :    . . . In the context you've got . . . richly labelling in the (singulate)
 2        (3.0)
 3 ( ):   Huh?
 4 J:     Mm mmm
 5 R:     I don't know whether there's an artifact
 6        (1.0)
 7 R:     But [ the
 8 J:        [ Check the other sections.
 9 R:     They're the same.
10 J:     They're the same?
11 R:     Uhhmm, noh,
12        (1.0)
13 R:     Noh, uh, but, uh, but basically the same, yes.
14 J:     There is something in- in th- in the singular cortex?
15 R:     There is something
16 J:     Well,
17        (1.0)
18 J:     That would make sense.
```

Following ethnomethodological standards, the transcription involves a minimal amount of cleaning up. Conversational analysis turns up regularities and organization not just in the most formal version of exchanges, but in their pauses, interruptions, modulations of voice, and so on. Here, for example, line (4) is taken as a weak "token of understanding" of (1), but given the long pause in (2) it is not taken as acceptance of the claim. Because J has held back acceptance, R raises the possibility of the labeling being an artifact. Following the demand in (8), that possibility is explored. Line (9) is an attempt to put it to rest, but is challenged. In (13), R asserts expertise in the matter, claiming that while the other slides are not *exactly* the same, they are *essentially* the same. Then, in (14) to (18), J lets the matter rest, accepting the original claim. Claims are made and then adjusted in the context of the conversation. Participants are attuned to each other's subtle cues, and react to them, adjusting their claims to increase agreement, though sometimes to highlight disagreement. In this particular case agreement is reached without reference to particularly technical details; R's expert judgment that the other sections are the same for the purposes at hand is allowed to stand on its own.

While conversational analysis might seem to be overly attuned to minutiae, it shows how the work of conversation contributes to the establishment of facts. Especially when researchers are engaged in collaborative projects, and need to come to agreements, conversations display how researchers decide on the nature of data. Klaus Amann and Karin Knorr Cetina (1990) investigated similar issues in a molecular biology laboratory. There, a key product was the electrophoresis gel, which when put on film shows indistinct light and dark bands that represent the lengths of different DNA fragments. Because these films are difficult to interpret, there is often a period of consultation during which their nature is typically resolved. Here is a sequence in which the researchers are performing what Amann and Knorr Cetina call "optical induction":

Jo: if you shift this parallely with the others, right, like that, that way, that way, this is nonetheless not running on the same level as this
Ea: no, this isn't on the same level, granted. I am not saying (it is), but I say/this is/say/
Mi: but if these run on the same level, this greatly suggests, doesn't it, that this is the probe
Ea: sure this is the probe. But then I also know that I've got a transcript which runs all the way through
Jo: but this can be this/this one here. That is this band here.
                              (Amann and Knorr Cetina 1990: 101)

The researchers here are deciding what they are looking at, drawing correspondences between different parts of the film, and between the film and the procedures that produced it. The result of such a conversation is that a complex array of light and dark bands is given an interpretation. It then becomes a defined piece of data to be drawn upon in later arguments.

Such conversations, which are often laborious and involve close interaction, cannot be part of published work. Other mechanisms must replace conversations in order to turn data into convincing evidence. For example, published images are very rarely the same as the images that are puzzled over in the laboratory. In the field that Amann and Knorr Cetina studied, published images were "carefully edited montages assembled from fragments of other images" (1990: 112). These montages could then highlight the orders that had been achieved. Artifacts – by-products of laboratory manipulations, not natural objects – and other aspects of the original films that were considered unclear, or irrelevant were trimmed from or otherwise de-emphasized in published photographs. The published fragments were juxtaposed to increase their meaningfulness, and "pointers" were added to push readers toward particular interpretations.

Lynch (1990) discusses similar issues under the rubric of *mathematization*. Images that are to serve as evidence in public contexts are (1) filtered, to show only a "limited range of visible qualities"; (2) made more uniform, so that judgments of sameness are projected on to the images; (3) upgraded, so that borders are more sharply marked; and (4) defined, by being visually coded and labeled. After such operations, the images become more like diagrams. Diagrams are not merely simplified representations, but are icons of "mathematical" forms to which the researchers want to refer: "the theoretical domain of pure structures and universal laws which a Galilean science treats as the foundation of order in the sensory world" (Lynch 1990: 162–3). While they may reflect the messy empirical domain, most untreated pictures of scientific objects are only poor indicators of the more clean and pure theoretical domain. Mathematized images help to form a connection between the empirical world and a world of orders and structures (see box 12.1); in this sense they are literary representations.

With all of this work, researchers eventually produce observations of the sort that can feed into the scientific analysis, observations that can form a basis of universal claims. Mention of the local work is generally not included in scientific publications, except insofar as it is describable in general methodological terms. Thus, the universal claims rest on observations produced by universal procedures, though the procedures are achieved locally.

Box 12.1   Mathematized lizards

In field ecology clean data are particularly difficult to obtain. Wolff-Michael Roth and G. Michael Bowen (1999) look at an ecological study of a lizard species, in which as many lizards as possible at a site were collected, marked, and variously measured. Analysis of the data was intended to establish a hypothesis about patterns of reproduction. Here data or observations are the products of scientific work at a particular stage in the research: they are the products that feed directly into mathematical and statistical analysis; they are already essentially mathematized versions of the lizards.

Collecting the lizards is itself a skill. In the terrain being studied, the animals are found under rocks, and thus researchers walk over their areas turning over rocks. The researchers chose not to turn over large piles of rocks, or rocks in clusters, because the noise that they made in so doing frightened away the lizards below and in the surrounding areas, sending them to more secure hiding places. Instead, they would turn over isolated rocks: their sampling technique was in this way, and in a number of others, attuned to the local difficulties of sampling. The researchers interpret general methodological rules when they apply them.

When the researchers attempted to take measurements of color, head width, tail length, sprinting speed, and so on, they again had to learn from handling the lizards how to make good observations. To measure the color, for example, the ecologists used a "Munsell Chart," a book of color samples. One researcher "had particular concerns about the accuracy of colour determination . . . and she was also worried about colour variability; she therefore set up conditions in the field laboratory that allow her to control variation. She constructed a special arena, a cardboard box closed on three sides by white walls, and installed a daylight incandescent lamp" (Roth and Bowen 1999: 749). Even then, matters were not straightforward. The researchers had to choose a consistent location on the lizards from which to read color, they had to learn how to simultaneously keep the lizards still and flip pages of the Munsell Chart back and forth, and so on.

Crystallization of Formal Accounts

In a widely read article, the eminent biologist Sir Peter Medawar (1963) asked, "Is the scientific paper a fraud?" His argument was that empirical scientific articles are written as if they were narratives of events, but are also written to be arguments. Since sequences of events do not typically form good arguments, narratives are adjusted to lead toward desired conclusions. Thus aspects of laboratory work like months of tinkering with experimental apparatuses, endless discussions about the nature of data, false starts, and abandoned hypotheses are not included in formal accounts of an experiment. And that is not even to mention such work as hiring and training research assistants, negotiating for money and laboratory space, responding to reviewers' comments, and so on.

Formal accounts provide a lens through which scientists and engineers see even their own work. In his study of the TEA-laser (chapter 9), Collins noted that after his informants got their apparatuses working, they immediately attributed problems to human error. The machines with which they had been tinkering changed from being complex and unruly to being relatively simple and orderly. Before they were working, each component was a potential source of trouble; afterwards the machines had a straightforward design based on some simple principles.

In the end, then, scientists and engineers assume that nature is orderly. Diagrams can stand in for pictures because diagrams represent a "mathematical" nature. Formal accounts can stand in for the weeks, months, or years experimenters spend getting their experiments to work because formal accounts represent the order of the experimental design. The work of the laboratory is work to make manifest the order of nature, and thus it cancels itself out from the self-image of science.

None of this is to say that formal rules and tools have a fictional status only. In addition to being goals they are also results. As results, these formal tools have powers of combination, that more messy and context-laden accounts do not have. It is these new powers that allow science and technology's successes to be more than isolated events or miracles.

Box 12.2   A sociology of the formal

In research on the computerization of record-keeping in a hospital, Marc Berg (1997) begins work on a "sociology of the formal." A computer at each hospital bed is fed information from various monitors, and nurses and doctors are expected to make regular entries. The result is an active database, which calculates patients' fluid balances and plots their status. The formal tool shapes the work done by nurses and doctors. The fields to be filled in pattern the nurses' and doctors' actions by demanding information. However, the computerized record is descended from paper records, and the paper records themselves reflect prior practices – though each incorporates changes based on new compromises of ideals, goals, and limitations. As a result, each stage in the development of the record changes practices only slightly. There is a co-evolution of the work done in the hospital and its representation in the formal records. The computer does not suddenly create a new, Taylorized regime. Rather, the work done by nurses and doctors is shifted by the presence of the computer. Similarly, the computer does not suddenly create a new, abstracted patient. Rather, the view of the patient is closely descended from the view already created by well-established practices.

But people work with computers, rather than simply have their work shaped by them. The nurses and doctors of Berg's study routinely correct the information on the computer screen, make fictional entries to allow their work to proceed, and otherwise creatively evade the limitations of the system.

The nurses and doctors (and no doubt the hospital administrators, as well) also take advantage of the new power of the formal tools. "The tool selects, deletes, summarizes; it adds, subtracts, multiplies. These simple subtasks have become so consequential because they interlock with a historically evolved, increasingly elaborate network of other subtasks" (Berg 1997: 427). Formal tools afford new opportunities for creating order, and opportunities for new orders.

Conclusions

The idea of formal objectivity runs against the demands of the messy world and the constant possibility of novel interpretations. So the question for this chapter was how to reconcile views of science that emphasize the strength of objective procedures in science and technology with humanist models that emphasize expertise.

ANT and other models that emphasize objectivity claim that science's pattern of successes can only be understood as built on objective or mechanical procedures. When we look closely at supposedly objective procedures, though, we find that they are shot through and through with human skill dealing with local contingencies. We should not see those procedures as produced by internal mechanical rules, then. Rather, objectivity is produced by human abilities to create order. We will thus always see expertise at work if we want to. But we can also abstract away from that expert work, trusting people to fit their actions to those objective procedures and to make formal tools capable of performing valuable work.

Science and technology can be seen as an extremely efficient producer of social order. As with any group of people, every time scientists and engineers come to agreement about an observation, an interpretation, a phrasing, or a theory, they are producing order. Though nature does not routinely force its regularity upon researchers, science and technology's orders typically come to be seen as the result of the order of nature. And that is why the work done by scientists and others to make the order of nature manifest is so interesting.

CHAPTER 13

Feminist S&TS and its Extensions

Can There Be Feminist Science and Technology?

Science and engineering are not fully open to participation by women, as we saw in chapter 4. Women face difficulties entering the education end of the pipeline, remaining there to build and maintain careers, and developing their reputation to gain status and prestige. As a result, they are underrepresented, especially outside of the life sciences. Some minority groups are similarly underrepresented, and people from outside a relatively few technically highly developed countries face even more substantial barriers.

What are the effects of those barriers on scientific knowledge and technological artifacts? What would result from removing those barriers? Chapter 4 focused purely on questions of equity, rooted in issues of justice and efficiency: discrimination is clearly unjust, and inefficient because it reduces the pool of potential contributors to science and technology. But in what ways would science and technology be qualitatively different if women were better represented? In what ways would science and technology be qualitatively different if feminist viewpoints were better represented? In what ways would science and technology be different if Western minorities, and non-Westerners, were better represented? This chapter sketches some relevant research, and some different answers, focusing on feminist science and technology studies (S&TS) – which has a rich tradition.

Until recently, it would have been straightforward to argue that, except in some small differences in focus, the content of a fully representative science and engineering would not be any different than it is. Scientific and technical arguments would be the same, and so, therefore, would be the knowledge and artifacts that they produce. Science should be teleological in the sense that scientific knowledge reflects, in as objective a way as possible, the structure of the natural world. Technology should be teleological in that it fills needs as efficiently as possible given the available resources and knowledge. However, given the argument of this book so far, science and technology are not such purely objective and teleological activities. What counts as knowledge, and what comes to be made, depend at least somewhat on local contexts. Therefore, we should expect that feminist science and technology would be different from science and technology as they are.

The Technoscientific Construction of Gender

Gender is the subject of a considerable amount of scientific study, and the object of a considerable amount of technological effort. For biologists and psychologists, sex-linked differences are important and interesting; they are, for example, central to the field of sociobiology. Whether those differences are studied in rats or in humans, differences between men and women represent at least part of the interest in the study. As a result, feminist S&TS has paid particular attention to biologists' studies of sex and gender – when it was not investigating and challenging the place of women in science and technology, early feminist S&TS was almost entirely devoted to questions about the scientific construction of gender. Books with titles such as *Women Look at Biology Looking at Women* (Hubbard, Henifin, and Fried 1979), *Alice through the Microscope* (Brighton Women and Science Group 1980), and *Science and Gender: A Critique of Biology and its Theories on Women* (Bleier 1984) made important contributions to understanding the scientific depiction of gender, with an eye to challenging it. Indeed, many of the most prominent critics of biology were themselves biologists, challenging the research of their colleagues with the aim of improving the empirical quality of the research.

Anne Fausto-Sterling's *Myths of Gender* (1985), for example, is a sustained attack on sex differences research, uncovering poor assumptions and unwarranted conclusions, showing how researchers fail to understand the social contexts that can produce gendered behavior, and challenging the ideological framework that supports sex difference research as a whole. Given the apparently large public interest in "biologizing" gender differences – and thus naturalizing and even legitimating them – critiques like those of Fausto-Sterling and many other feminist researchers have had some urgency. The size of their impact is difficult to gauge, especially since sex differences research continues to be popular both within science and on the pages of the popular media. However, at the least they have contributed to an increase in sophistication on the part of biologists and psychologists, aware that their work may be subject to some scrutiny. Fausto-Sterling's (e.g. 1993) controversial work on intersexed children, and how doctors intervene to place these children in one or another sex shows an even more direct attempt to reinforce established genders – and sexes.

Much of the study of the scientific construction of gender looks at how cultural assumptions are embedded in the language of biology. Emily Martin (1991), for example, explores common metaphors that describe the production and meeting of egg and sperm. Since the early twentieth century, the sperm has usually been depicted as active, and the egg as passive. The egg "does not move or journey, but passively 'is transported', 'is swept', or even 'drifts' along the fallopian tube. In contrast, sperm . . . 'deliver' their genes to the egg, 'activate the developmental program of the egg', and have a 'velocity' that is often remarked on" (Martin 1991: 489; see also Biology and Gender Study Group 1989). Then, when sperm meets egg, a similar active/passive vocabulary is used, the sperm "penetrating" a "waiting" egg. This active/passive vocabulary survives despite evidence that both egg and sperm act during fertilization. As Martin puts it, science has constructed a "romance" based on stereotypical roles, a romance that might reinforce those roles.

A number of more historical studies have made similar observations about depictions of males and females and their organs in biology, psychology, medicine, and anthropology. An interesting example that shows a surprising connection between science and gender is Londa Schiebinger's (1993) argument that mammals are called "mammals" because of the social and symbolic significance of breasts in eighteenth-century Europe – Schiebinger shows that breasts feature prominently in contemporary iconography, symbols of nurturing and motherhood. Linnaeus introduced *Mammalia* over the previous term *Quadrupedia* in 1758, in a context that included campaigns against wet-nursing. The label served to make breast-feeding a defining natural feature of humans (and other quadrupeds), and therefore served as an argument against hiring wet-nurses.

At the same time that science constructs images of gender, technology often embodies images of gender, and in so doing creates constraints on women. Reproductive technologies have built into them pictures of norms of sexual behavior, desires, and families (e.g. Ginsburg and Rapp 1995). Domestic technologies have built into them pictures of norms in households, standards, and divisions of labor (Cowan 1983). The built environment itself is structured around norms of divisions of labor and patterns of behavior – for example, aircraft cockpit design reflects images of users (Weber 1997). Such lessons are made abundantly clear in Judy Wajcman's survey, *Feminism Confronts Technology* (1991).

A classic study of technology and gender is Cynthia Cockburn's *Brothers* (1983), a study of conflicts over computerization in the London newspaper industry. Cockburn argues that choices about technologies are often the result of struggles between employers interested in reducing costs or increasing dominance, and skilled male workers interested in maintaining pay and status – particularly masculine status that comes from differentiating their work from women's work. Thus male industrial workers often want to work with heavier tools and machinery, making their work dependent on strength; alternatively, if it is possible they may be willing to move to more managerial positions or to be put in charge of machines (Cockburn 1985). Employers, however, are often interested in mechanization of tasks, either because it simply increases efficiency or because it allows them to hire less-skilled and lower-paid workers, often women. Mechanization and feminization, then, often go hand in hand. If Cockburn is right, then even industrial technologies are deeply gendered, and have an impact on gender structures.

Technologies are deeply political, because they enable and constrain action. Therefore, assumptions about gender roles that are built into technologies can, like assumptions about gender built into scientific theories, reinforce existing gender structures.

From Feminist Empiricism to Standpoint Theory

According to *feminist empiricists*, even though there are many instances of sexism in science, systematically applied scientific methods are sufficient to eliminate sexist work. In other words, feminist empiricists believe that it is possible to separate a purified science from the distorting effects of society. In some sense, feminist empiricism is the position on which other feminist approaches to the sciences are grounded. It occupies a primary place because it originates in the self-conception of the sciences; and feminist empiricist work began before work in and articulation of other feminist epistemological positions. At the same time, feminist scientists and researchers in S&TS have been so successful at identifying sexist assumptions in accepted scientific descriptions and theories that they have thrown doubt on the ability of at least the normal application of scientific methods to eliminate sexism (Harding 1986).

Helen Longino's *Science as Social Knowledge* (1990) is an attempt to understand how social ideology and the larger social context can play an important role in scientific inquiry, without giving up the idea that there can be some objectivity and room for specific criticism in science. Longino's goal is to show the role of values in science, without saying that science is so value-laden that there can be no foothold to criticize the justifications of specific claims. She focuses on background assumptions: what a fact is evidence for depends on what background assumptions are held. This allows people to agree on facts and yet disagree about the conclusions to be drawn from them. At the same time, which background assumptions people choose, and which ones they choose to question, will have a lot to do with social values. Feminist scientists of the past 20 years, with experience of sexism and anti-sexist movements, approached their subject matter with some different and novel presuppositions; the result has been a body of feminist science, especially in the biological sciences.

Although it might appear trivial, feminist empiricists have to explain why this critical work has been done primarily by women, and primarily by women sensitized to feminist issues. Any attempt to do so means rejecting purely atomistic ideas of scientists, in which they are for theoretical purposes identical, moved only by the laboratory data. There can be no general project of isolating science from distorting social assumptions, but only one of improving the social assumptions that science employs. Therefore, feminist empiricism undermines itself, needing at least a supplement to account for its own central cases.

Feminist empiricism's trajectory naturally takes it toward *standpoint theory* (or standpoint epistemology), a theory of the privilege that particular perspectives can generate. Nancy Hartsock, in her early articulation of feminist standpoint theory (1983), argues that the feminist standpoint is a *privileged* perspective, not merely *another* perspective. Other important (and different) versions of feminist standpoint theory are constructed by Dorothy Smith (1987), Hilary Rose (1986), and Sandra Harding (1991; see also Sismondo 1995).

The central argument of standpoint theory is that women have a distinctive experience, in the experience of sexual discrimination, from which to see gender relations as they really are. They are able to see aspects of discrimination that cannot be seen from the male perspective. This privileged position becomes fully available when women are active in trying to overturn male discrimination, for it is then that they necessarily see genders as non-natural, and unjust in their construction: "A standpoint is not simply an interested position (interested as bias) but is interested in the sense of being engaged. . . . A standpoint . . . carries with it the contention that there are some perspectives on society from which, however well-intentioned one may be, the real relations of humans with each other and with the natural world are not visible" (Hartsock 1983: 285). For the task of recognizing bias and discrimination against themselves, whether in scientific theories, science, or some larger society, women are uniquely positioned. More generally, people for whom social constraints are oppressive can more easily understand those constraints than can others.

Since S&TS argues that all science is a thoroughly social process, social locations that tend to make visible the effects of social structure should be valuable throughout the sciences. In particular, feminist scholars have argued that because scientific communities have lacked diversity, they have typically lacked some of the resources to better enable them to see aspects of sexism and androcentrism in scientific work. Therefore, standpoint theory, together with the recognition of the social character of knowledge, shows that to increase objectivity, communities or research and inquiry should be diverse, representative, and democratic.

From Difference Feminism to Anti-essentialism

What we might call *difference feminism* claims that there are masculine and feminine perspectives and styles of knowing, which may be correlated with the perspectives and styles of men and women. This is in contrast to standpoint theory's claim that social positions can offer particular insights. In S&TS the difference feminist position is probably most associated with Evelyn Fox Keller's *A Feeling for the Organism* (Keller 1983; also Keller 1985), a biography of Nobel prize-winning geneticist Barbara McClintock, but can also be prominently seen in such works as Carolyn Merchant's essay on reductionism and ecology, *The Death of Nature* (1980) and Sherry Turkle's studies of gendered approaches to and uses of computers (Turkle 1984; Turkle and Papert 1990). Outside of S&TS the position is probably best represented by the well-known arguments of Carol Gilligan (1982) that men's and women's approaches to ethical problems differ systematically.

The central claims of difference feminism in S&TS revolve around the idea that there are distinct gendered styles of scientific thought: masculine knowledge is characterized by reductionism, distanced objectivity, and a goal of technical control, and feminine knowledge by attention to relationships, an intimacy between observer and observed, and a goal of understanding. If this rough dichotomy is right, scientific knowledge itself is gendered.

We should not identify the masculine and feminine with men's knowledge and women's knowledge respectively, but instead identify a relationship between *gender* and scientific knowledge. For example, the scientist/Nature relation may be coded as a male/female relation, and a stereotypical key to the relation between the scientist and Nature is domination; this can be seen in uses of metaphors of domination and control, rape and marriage, where (male) scientists dominate, control, rape, or marry a (female) Nature (Keller 1985; Merchant 1980; Leiss 1972). Gender's place in science is at least partly defined by these relations, though women scientists can contribute to their disruption. Women scientists can be represented by Keller's McClintock: "a feeling for the organism" is an appropriate slogan for the way in which McClintock does genetics; rather than trying to dissect her organism she wants to understand what it is to be inside it. She says, "I found that the more I worked with them, the bigger and bigger [the chromosomes] got, and when I was really working with them I wasn't outside, I was down there. I was part of the system" (Keller 1985: 165).

The rough dichotomy that difference feminism posits has a particularly strong foothold in thinking about technology. According to Turkle, men's and women's approaches to computers can be roughly characterized in terms of a dichotomy between "hard" and "soft" mastery (Turkle 1984). Many engineers themselves employ a similar framework, understanding (masculine) technical prowess to be at the core of engineering (Hacker 1990), thereby downplaying (feminine) social aspects of the profession. Even researchers who are cautious about the value of the dichotomy find that it has some applicability. Knut Sørensen and co-workers (Sørensen 1992; Sørensen and Berg 1987) find both agreement and disagreement with such claims of gender divisions; although Sørensen finds few significant differences between male and female engineers with respect to their work, women tend to resist an eroticization of the technologies on which they are working.

In a useful discussion of these issues, Wendy Faulkner (2000) points out some interesting difficulties in the difference feminist project as applied to engineering. While gendered dualisms that describe approaches, styles, epistemological stances, or attitudes toward material are constant touchstones in thinking about engineering practice, they oversimplify the terrain. Faulkner draws attention to a number of related difficulties: in practice both sides of these dualisms are necessary, and "coexist in tension"; the two sides of these dualisms are not valued equally, and therefore aspects of technological practice falling under the less-valued half are downplayed; and these dualisms are often gendered in contradictory ways (Faulkner 2000: 760).

For example, the abstract/concrete contrast can be variously gendered. On the one hand, the ideology of engineering emphasizes concrete hands-on abilities, and when it does so those concrete abilities can be valued as masculine. On the other hand, engineering values the emotional detachment of mathematics, and when it does so abstract thinking can be valued as masculine. Men can be "hands-on tinkerers," or women can be perceived as engaging in a *bricolage* style of engineering. And in practice, it may be difficult to sort out such styles from the context of how the particular engineering cultures address particular problems.

The question about the importance of gendered dichotomies in practice is quite general. Companies in which people skills are more highly valued relative to technical skills tend to see more men in roles that are defined around those people skills. "Men are more likely to gravitate to those roles which carry higher status (or *vice versa*)" (Faulkner 2000: 764). In her own ethnographic work, Faulkner finds that women do not *report* the same level of fascination with technology as do men, even if in practice they show the same level of absorption in technicalities. So, gender may be more prominent in discourse than it is in practice.

In a direct critique of Keller's work, Evelleen Richards and John Schuster (1989) make some similar points. Richards and Schuster show that the gender structure of scientific practice is open to flexible interpretation. Looking at some variations on the story of Rosalind Franklin, whose empirical research was important to James Watson and Francis Crick's model of the structure of DNA, Richards and Schuster show that the same methods can be presented as feminine or masculine. What is gendered is method talk, not methods.

Such anti-essentialist challenges are tied in with movements in feminism more generally (e.g. Fraser and Nicholson 1990). One of the starting points for Donna Haraway's well-known essay, "A Manifesto for Cyborgs: Science, Technology, and Socialist Feminism in the 1980s" (1985) is the splintering of feminism in the early 1980s, when women of color made clear their frustrations with white feminism. For Haraway, the experiences and claims of women of color make clear that "woman" is not a completely natural political category. Their, and everybody's, identities are "fractured," structured by a multiplicity of causes – and thus the idea of discrete standpoints is open to serious question. For Haraway, feminism should cautiously embrace science and technology, in order to play a role in shaping it around local interests. As Haraway says, she would rather be a cyborg than a goddess.

The cyborg as a unit in social theory is Haraway's own invention, creatively adopted from science fiction, and emerging from a questioning of traditional political categories, from treating societies as information flows – that which transmits, receives, codes, and recodes information need not be human, but is a human/machine combination; it might be seen as "posthuman" in its rejection of the idea that the bounded individual human represents the important locus of thought and action (Hayles 1999). Cyborgs' postmodern origins join origins from the blurring of some traditional borders or boundaries – between organism and machine, between human and animal, and between physical and non-physical – accomplished by science and technology in the past hundred years. Cyborgs are in large part a result of our scientific society, providing another reason why it is not in their interests to be Luddites.

Gender, Sex, and Cultures of Science and Technology

As the field has grown, feminist S&TS has deepened the understanding of the places of gender in science and technology, and the relations of women to science and technology. Empirical topics continually arise: science, technology, and gender are central features of the modern world, and their interactions are various. Possibly, there are so many empirical topics that there is less room for synthetic analyses.

The places of women just outside the formal borders of science and technology have become interesting for what they show about women's informal contributions. The tradition of studying underrecognized women in science and technology is well established (e.g. Davis 1995; Schiebinger 1999), but it has also taken some novel turns aimed more at understanding cultures than simply at recovering contributions. To take one example, in a finely textured study, Deborah Harkness (1997) examines the "experimental household" of Elizabethan alchemist and mathematician John Dee – and his wife Jane. Jane Dee not only had to manage a household, but she had to manage her own location between Elizabethan society and natural philosophy. It was a location that was difficult to manage, because natural philosophy was done in the home, where Jane Dee had to act as a full partner to her husband, yet Elizabethan natural philosophy was chaotic, and its place not well defined.

Gender has become a more general topic, so that questions about masculinity by themselves have become interesting. These make, for example, science's being an almost exclusively male province potentially interesting. David Noble (1992) traces that male dominance back to the modeling of natural philosophical communities on monastic orders. But there are also local explanations. Mario Biagioli (1995) asks why the Accademia dei Lincei, an important organization of natural philosophers in Italy in the early seventeenth century, was established as an exclusively male order, complete with rules against various interactions with women, and encouragements of brotherly love. The justification could have been Platonic, based on an attempt to orient the natural philosophers toward the ideal world. Biagioli argues, however, that the reason is more simply based in the misogynistic vision of Federico Cesi, who funded and controlled the Accademia, and who saw women as a distraction from the business of finding out about the natural world.

Some of the extension of feminist studies is simply a result of focus on sites of key interest to feminists. Reproductive sciences and reproductive technologies are particularly prominent, because of the impact that they have on modern women. For example, Adele Clarke (1998) explores how in the twentieth century American reproductive sciences overcame a kind of illegitimacy because they dealt with sexuality and reproduction, socially forbidden topics, and because of their association with various frowned-upon movements. The low status of the reproductive sciences kept researchers and funding agencies from fully engaging with the biology of sex until after the Second World War, when powerful actors started supporting population control and birth control. Ethical and political issues that arise out of reproductive technologies have made them a prominent site for study within feminism generally (see, e.g., Burfoot 1999; Overall 1993). They have also been of anthropological interest: Charis Cussins's (1996) ethnographic study of infertility clinics explores how patients, eggs, sperm, equipment, are disciplined to make pregnancies and, more important for the clinics, data on achieved pregnancies (see also Franklin and Ragoné 1998; Ginsburg and Rapp 1995).

Some of the extension of feminist studies is the result of more incorporation of insights from the rest of S&TS into studies of sex differences. Nelly Oudshoorn's *Beyond the Natural Body* (1994) examines the development of sex hormone research, with a particular focus on how laboratory conditions had to be created to make sex hormones visible: "sex hormones are not just found in nature" (Oudshoorn 1994: 138). Instead, they are found in particular laboratory contexts, in which they can be made to act. They become seen as universal and natural because of "strategies of decontextualization" which hide the circumstances in which they can appear. Then, sex hormones go on to be used to help naturalize gender.

In S&TS, most studies of technology focus on the relatively "upstream" worlds of engineers, perhaps in interaction with scientists and/or entrepreneurs. There are relatively few detailed studies of the "downstream" worlds of users. This is despite the fact that interpretive flexibility is central to theoretical and programmatic views of technology (Pinch and Bijker 1987; Grint and Woolgar 1997). Some of the exceptions to this general picture are feminist studies of technology, which sometimes show contrasts between the understandings of technologies in the often male-dominated upstream worlds and the sometimes (depending on the technology) female-dominated downstream worlds. For example, Lisa Jean Moore's (1997) research on the use of "safer sex" technologies by prostitutes shows how users can give technologies novel meanings, even in the service of the ends for which they are made. In order to fit the contexts and culture in which prostitutes use it, an apparently clinical latex rubber glove can become sexy by being snapped on to hands in the right way. Or, it can be put to new uses by being cut and reconfigured. The users of safer sex technologies, then, actively produce those technologies' meanings and possibilities.

An interestingly different kind of novel use of technology is to make use of different ideologies of technology. Rachel Maines (2001) shows that the "electromechanical vibrator" was widely sold, and advertised in catalogues, between 1900 and 1930, despite the fact that masturbation was socially prohibited in this period, and was even thought to be a cause of hysteria. Maines argues that the vibrator could become acceptable for sale because of its association with professional instruments, because of the high value attached to electric appliances in general, and because electricity was seen as a healing agent. The modernity of technology, then, can be used to revalue objects and practices.

Race, Colonialism, and the Postcolonial Situation

Feminist insights into gender, science, and technology can easily be extended to address race, and colonial and postcolonial science. Most prominent in making that extension is Sandra Harding, whose edited collection on *The "Racial" Economy of Science* (1993) has been an important contribution to the field in its bringing together a wide variety of material; Harding has continued to focus on these issues (Harding 1998).

Some of the frameworks for studying the technoscientific construction of gender translate very easily to studies of race. A key starting point for the study of non-Western science is the identification of important developments that have taken place outside of Europe and North America, on some of which the Western scientific tradition depends. Martin Bernal's *Black Athena* (Bernal 1987) is a large-scale exploration of the African roots of European and Middle-Eastern culture, and of the forces that have served to obscure those roots – Bernal argues that the status of Ancient Greece as an almost isolated crucible of European intellectual development was created by classicists between the eighteenth and twentieth centuries. Such projects are analogous to the project of identifying women scientists whom standard histories of science have forgotten.

Race is as much constructed by science and technology as is gender. Indeed, Donna Haraway's *Primate Visions* (1989), intertwines discussions of gender, race, and location, showing how different narratives in primatology and in popularizations of primatology reflect important currents of thinking about men and women in Western societies, and also about relations to non-Western societies (see also Stepan 1986 on anthropology's metaphorical linking of race and gender). There are many studies of the anthropological construction of race, many of them important antiracist statements. Steven Jay Gould's *The Mismeasure of Man* (1981), for example, contains a number of excellent studies of attempts to show the superiority of Europeans over other races, and to mark boundaries between them. And in the practice of science we can see a different construction of the other: Warwick Anderson (1992) shows how researchers in tropical medicine constructed very different pictures of Filipinos and people of European descent.

Science can also police national boundaries. Agricultural research in the United States has on a number of occasions in the twentieth century established boundaries between "native" species and dangerous "immigrants." Washington's famous cherry trees were a gift from the Japanese government in 1912, but a precarious one. In a context that included anger and fear over (human) Japanese immigration, an earlier (1910) gift of two thousand cherry trees had been burned on the advice of the Bureau of Entomology, because they were variously infested (Pauly 1996). The later gift came along with strong arguments from Tokyo as to the health and safety of the trees. Some similar patterns can be seen in agriculturists' reactions to the kudzu vine (Lowney 1998).

Science and technology have served as important legitimations of imperialism and colonialism. Michael Adas, in *Machines as the Measure of Men* (1989; also Pyenson 1985) argues that when Europe started its colonial adventures the differences between European and non-European technological development were typically small enough that they did not provide a justification for colonialism; religion served that purpose. That changed, possibly partly because of colonialism. By the nineteenth century, European technological development, measured in terms of military power, the mechanical amplification of human labor, or in other ways, did form an explicit justification for colonialism. Technology was a symbol of Europe's modernity, and was something that Europeans could generously take to the rest of the world. Daniel Headrick (1988) argues, however, that the technology transfer from Europe to its colonies resulted more in underdevelopment than in the hoped-for industrialization. In broad terms, this theme has continued to be important in debates about the environmental impacts of Western agriculture and forestry (e.g. Nandy 1988).

Science could serve the same legitimating function as technology. Using a simple and elegant case, Kavita Philip (1995) argues that the transplantation of the cinchona tree (from whose bark quinine can be derived) from Peru to India via Kew Gardens in London was a colonial adventure performed in the name of science. The East India Company's Sir Clements Markham traveled to Peru's Caravaya forest to obtain specimens, where despite fierce opposition he managed to hire local cinchona experts to guide him to places where the trees grew, from which he took seeds and saplings. As Philip points out, Markham portrayed himself as serving science, though he had considerably less knowledge of cinchona than did his local informants. Opponents of his collecting activities were motivated by "patriotic zeal," and though his adventure was financed because of its importance to the British Empire, he was there to "conserve" the tree.

Postcolonial science can be vividly marked by its location. In an analysis by Itty Abraham (2000), an Indian "big science" project, the Giant Metrewave Radio Telescope, is described by its developers in terms of various features that make it the biggest of its kind. But it is also connected to nation-building projects, in that its site is justified as being the best such site in the world. While nationalist rhetorics often attach themselves to big science, even in and near centers, as in Canada (Gingras and Trépanier

1993), the rhetoric of postcolonial science can be seen as distinctly postcolonial. Abraham argues that the rhetoric around this telescope is designed around an Indian ambivalence to modernity: India simultaneously strives for modernity but does not want to simply mimic its former colonizer.

A key problem, then, for studies of postcolonial science is to understand the relationship between science in the less-developed and in more-developed worlds, between science in the "center" and the "periphery" (Moravcsik and Ziman 1975). Despite sizeable numbers of scientists, the less-developed world remains peripheral scientifically; it is "dependent" science, according to Susantha Goonatilake: "What is considered scientific knowledge in a dependent context is only that which has been made legitimate in the centre. It is then imitated in the periphery through the operation of pervasive dependent social and cultural mechanisms. . . . The fundamental and the basic core knowledge grows largely in the West and is transferred to developing countries in the context of a dependent intellectual relationship" (Goonatilake 1993: 260). Interestingly, that insight might also apply to peripheral locations within the "center," since people outside the core set of researchers in a field have little influence on the field's intellectual shape. Questions about the mechanisms of stratification (chapter 4) by location need more study.

CHAPTER 14

Rhetoric and Discourse

Rhetoric in Technical Domains?

Though there is no single scientific or technical style, we think of technical writing as typically dense, flat, and unemotional, precisely the opposite of rhetorical. Nonetheless, there is an art to writing in a factual and apparently artless style. The classical rhetorician Quintilian says, in a different context, "if an orator does command a certain art . . ., its highest expression will be in the concealment of its existence."

Wayne Booth says, "The author cannot choose whether to use rhetorical heightening. His only choice is of the kind of rhetoric he will use" (Booth 1961: 116). Science and technology studies (S&TS) adopts this expansive claim for rhetoric, for a straightforward reason: every piece of scientific writing or speech involves choices, and the different choices have different effects. The scientific journal article is an attempt to convince an audience of some fact or facts. That means that every choice its authors make in writing an article is a rhetorical choice. The arguments, how they are constructed, the language of their construction, key terms, references, tables, and diagrams, are all selected for their effects. What is more, the scientific writing process often involves interplay among multiple authors, reviewers, and editors, making at least some choices and their rationales open to investigation. These points are even recognized by authors who want to insist on the clear separation of rhetorical from epistemic issues (e.g. Kitcher 1991).

The fact that technical and scientific writing is not a transparent medium might be seen in the fact that dozens of manuals and textbooks on technical and scientific writing are published every year in English alone, the dominant scientific language. And scientific writing has not remained static. Charles Bazerman (1988) traces the history of the experimental report to show how it grew out of particular contexts, and was shaped by particular needs. When Isaac Newton wanted to communicate his theory of light, in his "New Theory of Light and Colours" of 1672, he recast a collection of observations to form a condensed narrative of an experiment. The single experiment was in some sense a fiction, a result of Newton's writing, but the narrative form was compelling in a way that a collection of observations and pieces of theoretical reasoning would not be. Newton was participating in already-established tropes of British science, and was refining the genre of the experimental report. Genre is by its nature not a stable category: Bazerman argues that each instance of it affects the genre of the experimental report. Thus it has changed and diversified considerably over the past 300 hundred years, becoming adapted to new journals, new disciplines, and new scientific tools. Nonetheless, as it was for Newton, today's report importantly has the potential to be a compelling narrative to make a particular point.

The Strength of Arguments

The point of most scientific discourse is ostensibly to establish facts. Though there are reasons to believe that important aspects of scientific discourse are not primarily about facts, or at least not about the facts they wear on their sleeves, establishing facts is clearly a central activity. Scientific articles are typically arguments for particular claims, experiments are usually presented as evidence for some more theoretical claim, and debates are about what is right.

What are the markers of a fact? According to Latour and Woolgar (1986 [1979]), the key markers are the lack of modality and history. Facts are presented without trace of their origins and without any subordination to doubt, belief, surprise, or even acceptance. The best-established scientific facts are not even presented as such, but are instead simply taken for granted by writers and researchers in their attempts to establish other facts. The art of positive scientific rhetoric, then, is the art of moving statements from heavily modalized positions to less-modalized positions. The art is to shift statements like "It has never been successfully demonstrated that melatonin does not inhibit LH" to "Since melatonin inhibits LH. . . ."

For Latour and Woolgar, key to this rhetorical art is operations on texts. In their study of a duel over the structure of thyrotropin releasing factor (or hormone) – with a Nobel Prize at stake – they show how articles from the two competing laboratories are filled with references to each other, but how those references involve skillful modalization and de-modalization (Latour and Woolgar 1986 [1979]: 132–3). Small markers of disbelief, questions, and assertions allow the two camps to create very different pictures of the state of the "literature" and the established facts.

For Latour (1987) the strength of an argument depends largely on the resources or *allies* that it brings to bear on the issue. Citations are one source of allies: a citation of another publication without modality, say ". . . (Locke 1992)," means that on that point David Locke will back the text up, and that in his 1992 book he provides evidence to support the text's claim. Latour shows how typical scientific arguments involve stacking allies in such a way that the reader feels isolated.

Latour's image is of an antagonistic relationship between writers and readers. The scientific article is a battle plan, or "one player's strokes in [a] tennis final" (Latour 1987: 46). Facts and references are arranged to respond in advance to an imaginary audience's determined challenges. Good articles lead readers to a point where they are forced to accept the writers' conclusions, despite their efforts to disagree. In a provocative turn, Latour says that when an article is sufficiently rhetorical, employing sufficient force, it receives its highest compliments: it is *logical* (Latour: 1987: 58).

For readers of scientific articles to challenge a claim requires breaking up alliances, for example by challenging cited evidence or by severing the connection between the cited evidence and the claim. This process typically takes them further and further away from the matter at hand, finding either weak points in the foundation or finding scientific commonplaces, facts too well established to be overturned. Thus while foundationalism is an inappropriate philosophical picture of science (box 2.2), it might be an appropriate picture of the rhetorical situation of scientific arguments.

There is another important element of Latour's rhetoric of science. The allies brought to bear on a position are not just cited facts, but include material objects, the objects manipulated in laboratories, found in the field, surveyed, and so on. These objects are represented in scientific writing, and in some sense stand behind it. Just as readers attempt to sever connections between cited and citing articles, they attempt to sever the connections between the objects represented and the representations. If this cannot be done conceptually, through clever questioning, readers may find themselves led back to laboratories where they can try to establish their own representations of objects. In this way, Latour's rhetoric of science makes space for the material world to play a role.

The Scope of Claims

Equally important to the strength of arguments is their scope, the audiences and situations that they can be expected to address. One well-cited text in the rhetoric of science is Greg Myers's *Writing Biology* (1990), a study of some related grant proposals, scientific articles, and popular articles (Myers 1995 has a similar analysis of patent applications). Myers had access to multiple drafts of each of these articles, along with reviewers' and editors' comments. He was able to piece together histories of their development, and show how these histories consist of a series of rhetorical responses to different contexts. In particular, considerations of the identities and interests of readers shaped the writing of the different texts Myers studied, producing works that were suited to promoting the authors' and reviewers' interests while catering to those of the readers.

Myers's central cases involve two articles that in their original forms made contentious claims. He draws on a framework developed by Trevor Pinch (1985) that contrasts differing levels of *externality* of claims. Arguments have higher levels of externality the wider the scope of their claims: an article that simply describes data typically has a very low level of externality, but one that uses those data to make universal claims about wide-ranging phenomena has a high level. Prestigious interdisciplinary journals such as *Science* and *Nature* accept only articles that make claims with high levels of externality, but reviewers are also attuned to issues of overgeneralization. In addition, reviewers want to see articles that make novel claims, but ones that address existing concerns. And they want to see journal pages used efficiently, so only important articles can be long ones. Thus articles must simultaneously make novel claims of wide scope and support those claims solidly, in tightly argued articles that address existing concerns of readers.

The review process centrally involved adjustments of the level and status of each of the claims made by the articles. In both of Myers's cases, articles were originally rejected by the prestigious journals to which they were submitted, were resubmitted, rejected again, and eventually found their way to more specialized journals. In both of his cases, the initial reasons for rejection centered on inappropriateness, because of the scope of the claims. In both of his cases, articles went through multiple drafts, ending in multiple articles addressing different audiences. The result of this process was the publication of scientific articles that made different claims than did the original submissions, and were substantially different in tone.

Rhetoric in Context

The rhetorical analyses mentioned so far are studies of scientific persuasion in general, even while they are attached to specific texts. But persuasion needs to be attuned to its actual audience, and for this reason rhetorical analysis often needs to pay attention to how the goals of individual writers interact with norms of scientific writing and speaking.

In the broadest terms, G. Nigel Gilbert and Michael Mulkay (1984) identify two *repertoires* that scientists use, in different circumstances. When discussing results or claims with which they agree, or when they are writing in formal contexts, scientists use an *empiricist* repertoire that emphasizes lines of empirical evidence and logical relations among facts: the empiricist repertoire justifies positions. When discussing results or claims with which they disagree, scientists use a *contingent* repertoire that emphasizes idiosyncratic causes of the results, and social or psychological pressures on the people holding those beliefs: the contingent repertoire explains, rather than justifies, positions. Since scientists move back and forth, Gilbert and Mulkay challenge the priority of the empiricist over the contingent repertoire, insisting that both have to be subject to discourse analysis.

Lawrence Prelli (1989) introduces the classical rhetorical notion of *topoi* or *topics* to discussions of science. Rhetorical topics are resources available in particular contexts. For example, he argues that something like Mertonian norms are topics available to science as a whole: to challenge the legitimacy of a scientist's work one might employ disinterestedness, displaying how extra-scientific interests line up with scientific judgments. There are also rhetorical topics specific to disciplines or methods; for example, there are recurring difficulties in attempts to apply the mathematical study of games, so a writer in this field has established topics to address (Sismondo 1997).

Rhetoricians might look at specific texts as more than examples (though also examples), seeing them as worthy of study in their own right. At least three prominent rhetoricians of science have scrutinized James Watson and Francis Crick's one-page article announcing their solution to the structure of DNA (Bazerman 1988; Gross 1990b; Prelli 1989). Stephen J. Gould and Richard Lewontin's anti-adaptationist article "The Spandrels of San Marco" is the subject of a book, representing about a dozen different rhetorical understandings of it (Selzer 1993).

When researchers are dealing with new or unfamiliar questions or phenomena, as opposed to answering a well-established question, they need to convince their readers that they are dealing with something real and worth paying attention to. The rhetorical task in these situations is to give presence to a question or phenomenon, to make it real and worthwhile in the minds of readers (Perelman and Olbrechts-Tyteca 1969). Alan Gross (1990a) shows, for example, the work to which taxonomists go to erect a new vertebrate species (the example is the establishment of a new hummingbird species). Not only do the taxonomists make arguments that their putative species is not already known, but they make their species salient in their readers' minds. Through pictures, charts, and a depth of observations, they create a unified phenomenon, available to readers.

Not all rhetoric is devised to establish facts. For example, Mario Biagioli's elegant study *Galileo, Courtier* (1993) looks at Galileo's work, and in particular at a number of his public disputes, in terms of Galileo's professional career: he successfully rose from being a poorly-paid, low-status mathematician to being a well-paid, high-status philosopher (or scientist). In seventeenth-century Italy, a key ingredient for a successful intellectual career was patronage. Thus Biagioli shows that Galileo's scientific work is often rhetorically organized to improve his standing with actual and potential patrons. Positions are articulated to increase not only Galileo's standing but that of his patrons. Even the very fact of the disputes is linked to patronage, in that witty and engaging disputes were a form of court entertainment, and contributed to the status of the court. Paula Findlen (e.g. 1993) makes similar points about the careers of other prominent scientists of that period; Jay Tribby (1994) shows the rhetorical use of experiments to support Tuscan nationalism in this period.

Reflexivity

The study of rhetorics and discourses has sometimes pushed S&TS in a reflexive direction. Recognizing the ways in which a piece of scientific or technical writing or speech is rhetorically constructed invites attention to the rhetorical construction of one's own writing (e.g. Ashmore 1989; Mulkay 1989). All of the rhetorical issues identified so far are also issues for work in S&TS. Data have to be given definition, propositions have to be modalized and demodalized, rhetorical allies have to be arranged, the scopes of claims need to be adjusted, appropriate topics have to be addressed, and problems have to given presence. Facts in S&TS are no less rhetorically constructed than are facts in the sciences or engineering. Reflexive approaches thus explore the construction of facts *per se*.

One of the ways in which rhetorics can be displayed is via *unconventional* or *new literary forms*. By writing in the form of a dialogue or play, competing viewpoints can be presented as coherent packages, and disputes can be left unresolved. Having a second, third, or more voices in a text provides a convenient tool for registering objections to the main line of argument, for indicating choices made by the author, or for displaying ways in which points made can be misleading (Mulkay 1985). Unconventional forms draw attention to the shapes that authors normally give their texts, effacing the temporality of research and writing, effacing the serendipity of topics and arguments, effacing the reasons and causes for texts appearing as they do. Unconventional forms, then, draw attention to the ways in which normal genres are conventional forms geared to certain purposes.

Reflexive approaches in S&TS have been the subject of some criticism. While they draw attention to conventional forms, and are often highly amusing to read, unconventional forms rarely change the relationship of the author and the reader significantly. Even while they appear to present challenges to the main arguments or claims, the challenges they present are in the control of the authors. Even while they appear to display temporality, serendipity, or background causes, they display these in the authors' terms. Finally, there is a tension between the attempt to establish a fact and the attempt to show the rhetorical construction of that fact, because the latter appears to delegitimate the former. Even though reflexive analyses typically show that S&TS is in exactly the same position as the fields it studies, those analyses appear to show weaknesses rather than strengths. Reflexive approaches, then, are very useful for learning about general processes of fact-construction, but do not by themselves solve any problems, and may create their own rhetorical problems (see Collins and Yearley 1992; Woolgar 1992, and Pinch 1993b; Woolgar 1993, for some interesting exchanges).

Incommensurability: Communicating among Social Worlds

Not all studies of communication are about rhetoric. Kuhn's strong claims about the incommensurability of scientific paradigms have raised questions about the extent to which people across boundaries can communicate. This question sometimes arises out of controversy studies, but its commonest home is in studies of disciplines.

It is in some senses trivial that disciplines (or smaller units, like specialties) are incommensurable. The work done by a molecular biologist is not obviously interesting or comprehensible to an evolutionary ecologist or a neuropathologist, although with some translation it can sometimes become so. The meaning of terms, ideas, and actions is connected to the cultures and practices from which they stem. However, people from different areas interact, and as a result science gains a degree of unity. We might ask, then, how interactions are made to work.

According to Peter Galison (1997), simplified languages allow parties to trade goods and services without concern for maintaining the integrity of practices. A *trading zone* is an area in which scientific and/or technical practices can fruitfully interact via these simplified languages or *pidgins*, without requiring full assimilation. Trading zones can develop at the interfaces of specialties, around the transfer of valuable goods from one to another. In trading zones, collaborations can be successful even if the cultures and practices that are brought together do not agree on problems or definitions.

The trading zone concept is flexible, perhaps overly so. To see this we might look at computer simulations. Because simulations are symbolic, they cannot belong to the realm of experiment, but because they produce unpredicted data, and because the details of their inner workings are opaque, they cannot belong to the realm of theory either. This can be an uncomfortable situation for computer modelers. "Caught between a machine life and a symbol life, computer programmers in physics risked becoming pariahs in the sense that many university physics departments found them neither fish nor fowl: they could successfully apply neither for theory positions nor for experimental jobs" (Galison 1997: 732; see also Dowling 1999; Kennefick 2000). From the point of view of modelers, their work may seem incommensurable with other work in the discipline. Nonetheless, to the extent that computer simulations are increasingly used, modelers must be learning to bridge their differences with theorists and experimenters, and other researchers are learning to speak some of the language of computer modeling. As such, computer simulation itself forms a trading zone between theory and experimentation. At different resolutions, simulations both require trading zones and are trading zones.

A different, but equally flexible, concept for understanding communication across barriers is Susan Leigh Star and James Griesemer's idea of *boundary objects* (Star and Griesemer 1989). In an historical study of interactions in Berkeley's Museum of Vertebrate Zoology, Star and Griesemer focus on objects, rather than languages. They want to understand how the very different social worlds of amateur collectors, professional scientists, philanthropists and administrators were bridged to allow productive collaborations. Each of those groups had very different visions of the museum, its goals, and the important work to be done. These differences result in various incommensurabilities among groups. However, objects can form bridges across boundaries, if they can serve as a focus of attention in different social worlds, and are robust enough to maintain their identities in those different worlds.

Star and Griesemer identify standardized records as among the key boundary objects that held together these different social worlds. Records of the specimens had different meanings for the different groups of actors, but each group could contribute to and use those records. The practices of each of the groups could continue intact, but the groups interacted via record-keeping. Boundary objects, then, allow for a certain amount of co-ordination of actions without large measures of translation.

The boundary object concept has been picked up and used in an enormous number of ways – it might even be seen as itself something of a boundary object bridging different approaches in S&TS. Even within the article in which they introduce the concept, Star and Griesemer present a number of different examples of boundary objects, including the zoology museum, the different animal species in the museum's scope, the state of California itself, and standardized records of specimens.

Why are there so many different boundary objects? The number and variety suggest that, despite some incommensurability across social boundaries, there is considerable communication. Though different groups may understand records of museum specimens quite differently, those records facilitate a form of communication. Though it may only be partial, there *is* an enormous amount of communication across specialties. Elke Duncker

(2001) looks at multidisciplinary research and concludes that communication is achieved via straightforward translation. Researchers come to understand what their colleagues in other disciplines know, and translate what they have to say into language that those colleagues can understand. Simultaneously, they listen to what other people have to say and read what other people write, attuned to differences in knowledge, assumptions, and focus. Thus concepts like pidgins, trading zones, and boundary objects, while they might be useful in particular situations, may overstate difficulties in communication. Incommensurability may not always be a very serious barrier to communication.

The divisions of the sciences result in disunity (see Dupré 1993; Galison and Stump 1996). A disunified science requires communication, perhaps in trading zones, or coordination, perhaps via boundary objects, so that its many fibers are in fact twisted around each other. Even while disunified, though, science hangs together and has a certain stability. How it does so remains an issue that merits investigation.

Metaphors and Politics

According to the positivist image, ideal scientific theories are axiomatic, mathematical structures that summarize and unify phenomena. In this picture, there is usually no important place for metaphors. They are viewed as rhetorical flourishes or, sometimes, as aids to discovery, but they are never essential to the cognitive content of theories. According to Mary Hesse (1966) and Donna Haraway (1976) almost every scientific framework depends upon one or a few key metaphors, and recent work in S&TS shows that scientific models and descriptions are at least replete with explicit metaphors and analogies. A number of works in the history of science and technology discuss particular metaphors, especially ones that have ideological value. We saw some of these in connection with feminist research (chapter 13); box 14.1 provides a few more examples.

Box 14.1   Some metaphors in science

Philip Mirowski (1989) puts the study of metaphor to use in a critique of the theory of value in economics. Neo-classical economics appropriated the formalisms of the contemporary physics of energy, resulting in a metaphorical connection between energy and value. Nineteenth-century physics, then, shaped much twentieth-century economic research, in ways and directions of which the researchers have been largely unaware. That lack of awareness, Mirowski argues, has contributed to confusions, both because economists have not accepted the consequences of their metaphors and because they have been unable to see alternatives. Mirowski suggests, then, that economists examine new metaphors or models of value, some of which might lead to new solutions to long-standing problems.

Donna Haraway (1976) also uses the study of metaphors in a constructive critique, attempting to revive a form of non-reductionism in biology. Haraway follows Hesse in arguing that metaphors are necessary to science because of their fertility, and her book explores whether such a position can fit with a Kuhnian model of science in which fertile metaphors are one component of paradigms. She studies the effects of a set of metaphors connected with organicism, an anti-reductionist position in developmental biology. By looking at the work of three important and interesting twentieth-century figures she argues that the metaphors of organicism – crystals, fabrics, and fields – provide some consistent directions of thought and work, and thus might be thought of as contributing to a loosely structured paradigm, that could be redeveloped. Haraway's recent work continues to focus on metaphors in biology, particularly on metaphorical relations between biological images and popular understandings of race, class, and gender (see, e.g., Haraway 1989, 1991).

The uses and effects of figurative scientific language are the focus of most of the essays in Evelyn Fox Keller's *Secrets of Life, Secrets of Death* (1992), essays which range widely over such topics as population genetics, the "competition" metaphor in ecology, and the Manhattan Project. The title essay juxtaposes Watson and Crick's assault on the "secret of life" and commonly used metaphors of paternity in descriptions of physicists who worked on the atomic bomb. The metaphors – some of them embedded in actions rather than being wholly linguistic – around these different secrets can be read through the lens of gender, affecting the symbolic politics of life and reproduction.

In thinking about science, ideology is often contrasted with purity: scientific knowledge is ideological when it has been distorted by issues of power, and pure when it has been unaffected. If sociological and political critics of science are right, then there is no such thing as pure science in this sense. For some analysts of science (e.g. Gergen 1986; Jones 1982) the ubiquity of metaphor even raises questions about science's claims to represent reality accurately: if metaphors are so prevalent, how can we say that scientific theories describe reality rather than construct frameworks? In particular, how can inquiry influenced by metaphors current in the wider culture, as so much of science is, be taken as representing?

Why are there so many metaphors? Metaphors in science are crucial as heuristic and conceptual tools (e.g. Hoffman 1985; Nersessian 1988), and often serve important descriptive and referential functions (e.g. Ackermann 1985; Cummiskey 1992). The ubiquity of metaphor and analogy in the sciences can be taken as evidence that literal language lacks the resources for easy applicability to new realms (Hoffmann and Leibowitz 1991). Metaphors can define research programs rich with questions, insights, and agendas for research (Boyd 1979). They can become so rich that they become invisible – Lily Kay (1995) argues that the dominant metaphor of genes as information was in fact a contingent development, though it is almost impossible to think of genetics except in informational terms.

Theories and models are abstractions, approximating away from the truth (chapter 15). Too-tight correspondence is something to escape from. At the theoretical level, scientists aim at elucidating the structures of material things. But abstractions have to take place within a framework, in the form of a lens through which to choose elements to abstract. Metaphors can provide such a lens, allowing ideology and truth to coexist.

Conclusions

Epistemic issues are simultaneously issues about persuasion. Even if our focus is narrowly on the production of scientific and technical knowledge, then, the study of rhetoric is crucial to understanding what people believe and how they believe it. Once that is granted, there are many different approaches to understanding persuasion, from Latour's picture of persuasion as achieved by a kind of rhetorical force to Prelli's picture of writers addressing disciplinary topics.

There is also more to science and technology than the direct production of knowledge. Scientific and technical writing is also tied to a variety of contexts, including professional ones for scientists and engineers, questions of the legitimacy of approaches and methods, and broad ideological contexts. Often these contexts can be made visible only by the study of rhetoric, showing the metaphorical connections between discourses, goals, and ideologies.

CHAPTER 15

The Unnaturalness of Science and Technology

The Status of Experiments

Since experiments came into their own as sources of knowledge in the seventeenth century, experiments have tended to replace field observations where possible. This is not an obvious development. Considering the unnaturalness of experimental observations, and their inaccessibility to other interested researchers, experimentation looks like a fragile way of obtaining knowledge about the natural world. However, science has largely adopted an "interventionist" approach to research (Hacking 1983). Thus experiments have been the focus of more attention in science and technology studies (S&TS) than have observations in the field (for a few exceptions, see Clark and Westrum 1987; Bowker 1994; Latour 1999).

Most understandings of science view experiments as ways of deciding among theories. In the falsificationist view (chapter 1), for example, theories are imaginative creations that stand or fall depending upon observations, typically laboratory observations. Theories have an independent value that experiments do not have. Thus in most scientific hierarchies theories sit above experiments, and theorists sit above experimenters. Attention to laboratories, however, has started to erode the theory/experiment hierarchy in S&TS's view of science, if not in science itself. There are a number of reasons for this.

"Experiments have lives of their own," says philosopher Ian Hacking. Many experiments are entirely motivated and shaped by theoretical concerns. But at least some experiments are unmotivated by anything worthy of the name "theory." Hacking quotes from Humphry Davy's 1812 textbook, in which Davy provides an example of what should count as an experiment: "Let a wine glass filled with water be inverted over the Conferva [an algae], the air will collect in the upper part of the glass, and when the glass is filled with air, it may be closed by the hand, placed in its usual position, and an inflamed taper introduced into it; the taper will burn with more brilliancy than in the atmosphere" (quoted in Hacking 1983: 153). While Davy's actions in his example are shaped by ideas, there is no important sense in which they are the products of a discrete theory or set of theories. And Davy certainly does not set out to test any particular theory.

Once we accept that there can be experiments unrelated to theory, must we accept that experimentation is unrelated to theory? Surely, even experiments like Davy's are justified in terms of their potential contribution to theory? This claim is certainly right, but we should note that it amounts to little more than a reiteration of the theory/experiment hierarchy. We can read Hacking's slogan that experiments have lives of their own as saying that the hierarchy leaves space for a relatively autonomous activity of experimentation.

One of the innovations of S&TS has been to look at science as work. Particularly, those members of the field most influenced by the symbolic-interactionist approach to sociology have emphasized how science consists of interacting social worlds (e.g. Star 1992; Fujimura 1988; Casper and Clarke 1998). Participants are invested in their worlds, and attempt to ensure those worlds' continuance and autonomy. They strive to continue their own work and maintain identities. Looked at from the perspective of work, experimentation and theorizing are not hierarchically related, but are different (broad) modes of scientific work.

Knorr Cetina says that "theories adopt a peculiarly 'atheoretical' character in the laboratory" (1981: 4). Her ethnographic observations lead her to adopt a picture of laboratory science as aiming toward success rather than truth. That is, the important goals of laboratory researchers are working experiments, reliable procedures, or desired products, rather than the acceptance or rejection of a theory. Knowledge is created in the laboratory, but "know-how" is primary, and "knowledge that" is largely a reframing of laboratory successes for particular audiences and purposes. In the laboratory theories take a back seat to more mundane interpretations, attempts to understand particular events or phenomena.

Finally, does theory contribute more to technology than experiment? It is not at all clear whether one is more important than the other. As has been suggested already, theories have to be supplemented to be applicable. Students of technology have pointed out that scientific knowledge is often inadequate for the purposes of engineers, and engineers often have to create their own knowledge (e.g. Vincenti 1990). The standard picture of technology as wholly dependent on science for its knowledge is flawed. Once we recognize that flaw, it is unobvious whether theoretical knowledge or experimental know-how provides more help to engineers. Again, S&TS suggests that experimentation is not simply subordinate to theorizing.

Local Knowledge and Delocalization

Experiments are rarely isolated entities, but are rather connected to others that employ similar patterns, tools, techniques, and subject matters. One way of grouping like experiments is by the notion of an *experimental system*, a concept used by Hans-Jörg Rheinberger (1997). Rheinberger points out that teams often devote tremendous energy to developing a combination of tools and techniques with which to run any number of varying studies. For example, the "Fly-group" of geneticists led by T. H. Morgan in the early decades of the twentieth century developed *Drosophila* into a tool for studying genetics (Kohler 1994). To do so they bred the fly to better adapt it to laboratory conditions, and they developed techniques for further breeding and observing it. On the basis of that experimental system a minor industry in *Drosophila* genetics emerged. Similar stories can be told about the Wistar rat (Clause 1993), the laboratory mouse (Rader 1998), and other standardized laboratory organisms.

As we saw from the discussion in chapter 9, laboratory work does not proceed in a purely systematic fashion. Plans from above run into local resistance, which is also accommodated locally. Materials do not behave as they are supposed to, equipment does not function smoothly, and it is often simply difficult to configure everything so that it resembles the plan. The laboratory contains local idiosyncrasies that become central to day-to-day research. Researchers face a broad range of idiosyncratic issues, from employment regulations – which might "prohibit testing after 4:30 p.m. or at weekends" – to problems with materials – "the big variability is getting the raw material. We have never been able to get the same raw material again. . . . You have to scratch yourself in the same place every time you play, and everything has to be the same, or else the accounts are meaningless" (quoted in Knorr Cetina 1981: 38).

Looked at in this way, the problem of replication is a problem of delocalization. An experimental result occurs in a specific time and place. Scientific knowledge tends to be valued, though, the more it transcends time and place. But to delocalize or generalize a fact is a difficult task, in which one has to identify what it is that is to be generalized, if anything. Knowing that dioxin was strongly linked to cancer in 20 laboratory rats, can one conclude that dioxin is carcinogenic in all mammals, in all rodents, in all laboratory rats, in all well-fed, under-stimulated animals . . .? What class of things did the 20 laboratory rats represent?

Trevor Pinch (1985) argues that the success of empirical work can be measured in the generality of the conclusions that are allowed to be drawn from it, or, in his more precise term, the *externality* (from the particular set-up of the study) of conclusions. "Splodges on the graph were observed" and "solar neutrinos were observed" are extraordinarily different statements of observations, though they might refer to the same event. The former is not likely to be challenged, but by itself has no importance. The latter stands a much higher chance of being challenged and may be of importance or interest. The goal of the researchers is to push their claims to as high a level of externality as they safely can, escaping the local idiosyncrasies of their work as much as possible.

The Unnaturalness of Experimental Knowledge

Scientists construct systems in order to escape the messiness of nature, to study controlled, cleaned, and purified phenomena, about which models can be more easily made. Experimentalists can know much more about artificial phenomena than messy nature. In a strict sense they rarely study nature. As Karin Knorr Cetina says, nature is systematically excluded from the lab (Knorr Cetina 1981; Hacking 1983; Latour and Woolgar 1986 [1979] – Radder 1993 challenges the importance of this).

Experimental subjects and the contexts in which they are placed may be entirely artificial, constructed or fabricated, making our knowledge about them not knowledge about a mind-independent world. The world of experiments, both material and thought experiments, can exist in the absence of naturalness – scientists can construct and investigate ever more simple or complex systems, according to their interests. Experimental knowledge is not knowledge directly about an independent reality, but about a reality apparently constructed by experimenters.

Put differently, experimental knowledge takes an enormous amount of work. As we have seen, it takes work to set up experimental systems that perform relatively reliably. Those systems are thus scientists' creations, having been purified and organized so that they do not behave as chaotically as nature, so that they do what the researchers want them to do. But experimental systems are rarely perfectly reliable, in the sense of responding well to every situation or set of inputs that experimenters wish to explore, so it also takes work to continually maintain the order of the laboratory. As we saw in the previous chapter, scientists attribute the order that they discover to nature, and the disorder to local idiosyncrasies or their own inadequate control of circumstances. But there is a case for attributing science's ordering of nature not to nature but to scientific work itself. The last section of this chapter explores this theme very briefly.

The artificiality of experimentation was one of the concerns that many natural philosophers of the seventeenth century had about its legitimacy. According to the scholastic ideal that dominated knowledge of the early seventeenth century, the natural phenomena that scientific knowledge studied were supposed to be well established or readily observable regularities, like the tides, the weather, or biological cycles (Dear 1995; Shapin and Schaffer 1985). Experiments could be demonstrative devices, but their artificiality prevented them from standing in for nature, and prevented them from establishing phenomena. Daniel Garber (1995) argues that to become legitimated experiments had to produce social facts as opposed to individual ones, and from that point display natural facts. Before they could become revealing of nature, experiments had to become seen as potentially replicable by all competent experimenters, thereby being universalized. Peter Dear (1995) argues that such concerns were addressed through the creation of an analogy between mathematical constructions and material ones: mathematical constructions enable learning about mathematical structures in the same way that material constructions enable learning about (natural) material structures. Particular places and spaces that served as laboratories contributed to the legitimacy of experiment – for example, the location of laboratories within the homes of English gentlemen helped establish trust in experimenters and the phenomena they displayed (Shapin 1988). And in a study of one prominent experimenter, Steven Shapin and Simon Schaffer (1985) argue that Robert Boyle's detailed writing allowed for *virtual witnessing* of his experiments with air pumps. This helped to overcome the difficult burden of arguing that they were not idiosyncratic, private events, but publicly available reflections of natural regularities.

The Unnaturalness of Theoretical Knowledge

There are some parallels to experimentation's artificiality in the world of theory. Although we often think of theories and models as straightforwardly about the material world, describing its states and properties, they are at best idealizations of or abstractions from the real world that they purport to represent. This point is at least as old as Plato's argument that mathematics must describe a world of forms that are prior to material reality. But it has also been made forcefully more recently. As Nancy Cartwright (1983), Wladyslaw Krajewski (1977) and others have pointed out, scientific theories and laws are *ceteris paribus* statements: they do not tell us how the natural world behaves in its natural states, but rather that a set of objects with such and such properties, and only those properties, would behave like so. By science's own standards, its best theories and laws rarely even apply exactly to the contrived circumstances of the laboratory.

Cartwright's (1983) examination of physical laws provides a number of examples of the artificiality of theories and laws (see also box 15.1).

She argues that physical laws apply only when everything is just right. To take a simple example, Newton's law of gravitation is: $F = Gm1m2/r2$. This states that the force between two objects is directly proportional to the product of their masses. But this is only right if the two objects have no charge, for otherwise there will be an electrical force between them, following Coulomb's law, which is similar in form to Newton's. We can correct these two laws by combining them, but at the expense of their usefulness in the spheres in which they are traditionally used. Or we can divide the single force acting between two bodies into two components, a gravitational force and an electrical one. Cartwright concludes that "we can preserve the truth of Coulomb's law and the law of gravitation by making them about something other than the facts" (Cartwright: 1983: 61).

Box 15.1   Idealized islands, idealized species

R. H. MacArthur and E. O. Wilson's (e.g. 1967) theory of island bio-geography (IB) predicts the number of species (within some taxonomic group) that will be found on an island, on the assumption that immigration and local extinction will eventually balance each other out. The level at which they balance depends upon the size of the island (because small islands have higher extinction rates), and how far it is from sources of colonization (because distant islands receive fewer immigrants). The theory was an important one in ecology, and the debate around it shows conflicting views of abstraction and idealization (Sismondo 2000).

IB was often viewed as obviously right at a high level of abstraction. To most theoreticians the truth of IB was not in question, even though its predictions were not very good, and even though it made assumptions that are not strictly true. To many empirically minded ecologists the situation was the reverse. The data did not support IB, and the theory made key assumptions that were too incorrect to accept: it assumed that all the species in question were essentially identical, and that all the islands in question were essentially identical except for size and location.

In the end, the debate came down to one over the relative status of theoretical and particular explanations. To the extent that ecologists were able to create particular explanations, and interested in creating them, they found IB too abstract to count as possibly true. To take a popular example, according to the theory the extinction of the dodo was made more likely by its being located only on small islands isolated in the middle of the Indian Ocean. But while the theory explains this extinction, it does so only in an unsatisfying way. The explanation

> discards too much information, information about, for example, the birds' naiveté, the explosion of human travel since the sixteenth century, and the voraciousness of modern hunters. In historical terms we know what happened to the dodo, and to many minds the theoretical terms offered by IB seem pale in comparison. To empirically minded ecologists, then, nature is the collection of particular events. In the process of investigating that nature they get their boots dirty or otherwise immerse themselves in data, and IB fails because it does not attend closely enough to those particulars.

For explanatory purposes, scientists want general laws that are widely applicable, even if that is at the expense of their being exactly applicable in any real case. Thus they move the discussion away from the straightforward facts of nature, making their theories about underlying structures of nature. Like experimental knowledge, theory is about cleansed and purified phenomena, abstractions away from the truth. Theories, models, and even many isolated facts do not mirror nature in its raw form, but instead may describe particular facets or hidden structures, or may be attempts to depict particular artificial phenomena.

A Link to Technology?

As we saw in chapter 8, it is not obvious that there should be a close connection between science and technology, and in fact the two stand somewhat further apart than many people assume. At the same time, science contributes substantially to technology, and the two seem increasingly difficult to disentangle (Latour 1987; Gibbons et al. 1994; Stokes 1997). The unnaturalness of experiment suggests a reason why science and technology are as linked as they are.

If experimental science provides knowledge about artificial realities, it provides knowledge about the structure of what can be done. Experimenters make systems in the laboratory, and investigate what those systems can do. Those systems may be very informative about the order of nature, but at root they are artificial systems. Engineers, too, create artificial systems, not usually with the goal of finding out about the natural world, but with the goal of doing some concrete work.

At the same time, given the studies of the gaps between science and nature, obvious questions arise about the possibility of materially bridging those gaps. How, if science does not describe nature, is science applied to nature? So, at the same time that it suggests a context for understanding the closeness of science and technology, the unnaturalness of both experimental and theoretical knowledge suggests a problem for understanding the *success* of applied science.

Box 15.2   Genetically modified organisms

Nature, of course, has immense symbolic value. Technological controversies can easily become controversies about the acceptable limits of the unnatural. Debates over anti-depressant drugs, new reproductive technologies, and genetically modified foods often revolve around the need for natural emotions, childbirth, and food.

Genetically modified (GM) foods are at the center of the debate over genetic engineering more generally. The number and variety of highly public issues that critics have raised is astonishing, suggesting deep-seated aversions to the idea of GM foods. Could GM foods provoke unexpected allergic reactions? Could foreign gene sequences get transferred to microorganisms in people's intestinal tracts? Do GM foods contain inedible and therefore harmful proteins? Will dangerous genes spread when GM crops hybridize? Will GM foods threaten the food chain when insects eat them? Will insecticides produced by GM plants threaten non-targeted insects? Will herbicides used on herbicide-resistant crops threaten non-targeted plants? Will herbicide use increase as a result of herbicide-resistant crops? Do GM organisms threaten biodiversity? Will GM crops simply increase the dependence of farmers on agribusiness? Will they lead to more consolidation of farms? – Proponents have, unsurprisingly, made an equally large number of positive claims for GM foods.

Anti-GM alliances are relatively unified, even when they involve somewhat contradictory positions, again suggesting deep-seated concerns. Pro-GM writers invoke consumer rights when GM foods are at issue, and anti-GM writers invoke consumer rights when the *labeling* of GM foods is at issue; similarly pro-GM writers emphasize the flexibility and uncertainty of knowledge behind labels, while anti-GM writers emphasize the flexibility and uncertainty of knowledge behind the technologies themselves (Klintman 2002).

Within scientific and technical communities, the loud public debates have their effects. To take one from many possible examples, when researcher Arpad Pusztai announced on television that GM potatoes had negative effects on rats' intestines, he was roundly criticized, and lost his job, for having circumvented peer review. An article on the study was eventually published by the medical journal *The Lancet*, but before it appeared the British Royal Society had already publicly pronounced it flawed, and the journal as biased. This provoked counter-charges from the editor of *The Lancet*: the Royal Society, he claimed, had staked its financial future on partnerships with industry, and as a result it and its members had been captured by industry interests. Because GM controversies are so public, scientific interventions often provoke strong normative criticism, for being interested or otherwise violating the scientific ethos. Scientific realms, then, are not immune to concerns over the technological encroachment on nature.

The Order of Nature?

What is scientific knowledge about? It is about all types of things, from macro-evolutionary tendencies to genetic mechanisms, from black holes to superstrings, from the global climate to turbulence. We might be inclined to say that scientific knowledge is about natural objects and processes, and also a few particularly interesting artifacts. At the same time, most of the tools that produce scientific knowledge place it one or more removes from such objects and processes. The best scientific knowledge does not straightforwardly consist of truths about the natural world, but of other truths.

When psychologists try to understand the structure of learning they typically turn to experiments on rats in mazes, or on people in constrained circumstances, but they more rarely try to study the everyday learning that people do in complex ordinary environments. Knowledge stemming from experiments most directly describes situations that are distinctly non-natural, standing apart from nature in their purity and artificiality. Experimenters make sure that their inputs are refined enough to be revealing, and use those inputs to create relationships not found outside the laboratory. They learn about something that they see as more basic than the immediate world. Similarly, science's best theories don't describe the natural things we observe around us, or even the invisible ones that can be detected with instruments more subtle than our senses. Instead, those theories describe either idealizations or other kinds of fundamental structures. Few basic theories literally or accurately describe anything material, according to science's own standards. And computer simulations and mathematical models create new symbolic worlds that are supposed to run parallel to the more familiar ones, but are in many ways distinct: simulations are typically explorations of what would happen given specific assumptions and starting conditions, not what does happen. Even field science necessarily creates abstractions from the natural world, depending upon systems of classification, sampling, and ordering; in these ways it can be seen to project laboratories on to the field.

Most scientific knowledge both is and is not universal. It is universal in the sense that in its artificiality and abstraction it is not firmly rooted to particular locations. Theoretical knowledge is about idealized worlds; laboratory knowledge is created so that it can be decontextualized, moved from place to place with relative ease. Scientific knowledge is not universal in the sense that its immediate scope is limited to the artificial and abstract domains from which it comes, though there is always a possibility of its extension. Latour has an analogy that is useful in this context:

> When people say that knowledge is "universally true," we must understand that it is like railroads, which are found everywhere in the world but only to a limited extent. To shift to claiming that locomotives can move beyond their narrow, and expensive rails is another matter. Yet magicians try to dazzle us with "universal laws" which they claim to be valid even in the gaps between the networks.   (Latour 1988: 226)

Locomotives are very powerful, but they can only run on rails. Scientific knowledge is similarly powerful, but when it leaves its ideal and artificial environments it can quickly become weak.

Very roughly, we can understand the objects of scientific knowledge in one of two ways:

1   In constructivist terms – they can be seen as constructions of the researchers, as those researchers transform disorderly nature into orderly artifacts. Theoretical and laboratory tinkering shapes nature so that it can be understood and used. Order is imposed upon nature by science.
2   In realist terms – they can be seen as revealing a deeper order, which is absent in surface manifestations of nature. Scientists tinker to reveal structure, not to impose it. Science is an activity that discovers worlds that lie beneath, or are embedded in our ordinary one.

For the most part, S&TS adopts constructivism over realism. However, in so doing the majority of scholars in S&TS have not adopted a magical or fanciful view. Rather, they believe that the mundane material actions of scientists in the laboratory produce new objects, not pre-existing ones. Sometimes they say something stronger than that, but not without thought or good reasons.

With some considerable philosophical work, it is possible that these positions can be reconciled. That would take us outside the scope of this book. This chapter leaves the question hanging, then, though it should show that the question is a genuine one.

CHAPTER 16

Expertise and the Public Understanding of Science

The Shape of Popular Science and Technology

We have seen that scientific and engineering research is textured at the local level, that it is shaped by professional cultures and interplays of interests, and that its claims and products result from thoroughly social processes. This is very unlike common images. It is very unlike, for example, the views of science and technology with which this book began. And it is very unlike the views one finds in popularizations. Why is this?

Although science and technology studies (S&TS) does not provide strong grounds for skepticism, the humanization of science and technology clearly undermines some standard sources of scientific and technical authority. We cannot say, for example, that scientific knowledge *simply* reflects nature, or even that it arrives from nature as the result of a perfectly rigid series of steps, so its authority cannot stem from nature alone. We cannot say that technologies *simply* unfold naturally and inevitably, and so the authority of engineers cannot stem from a too-easy narrative of progress. Claims about the constructedness of knowledge and technologies do come up in science and engineering, but only in very specific contexts. For example, they come up when scientists and engineers use intellectual property law to defend their interests; then they need to argue for the constructedness of their knowledge in order to assert their creative contribution to it (McSherry 2001; Packer and Webster 1996). More commonly, though, such claims come up when disputants in a controversy use the "contingent repertoire" to delegitimate their opponents – they tend to appear to diminish authority. For this reason the observations made in S&TS look threatening to scientists and engineers.

We can imagine that journalism inflected by S&TS would not be popular with scientists. According to Dorothy Nelkin (1995) and Christopher Dornan (1990), science journalists are very closely allied with scientists. All journalists depend upon their contacts for timely information, and science journalists are no exception. They thus participate in both informal and formal networks. Some important scientific journals, like *Nature, Science*, and the *New England Journal of Medicine*, send out advance copies of articles to select writers, on the condition that those writers not publish anything about those articles before the journal does (Kiernan 1997); that can allow a science writer to have a story, complete with interviews and quotes, prepared and waiting, and it allows the journals timely publicity. In addition, to an extent that other journalists do not, science writers depend on their contacts for accurate facts and background information. Science is often esoteric and difficult to understand, and accuracy is a key value in writing about science, which amplifies the dependence on contacts. It is possible, then, that science reporting in newspapers and magazines, and science writing in books more generally, is shaped by these dependencies.

Perhaps most importantly, the style of science journalism that has developed emphasizes findings and their importance, not processes (e.g. Gregory and Miller 1998). Newspaper and other editors are interested in stories that convey excitement. Those stories tend to be about the definitive – at least at the top of the story – discovery of the biggest, smallest, or most fundamental of things. Doubts, questions, caveats, and qualifications are downplayed. Thus most readers are left with the impression of one or a very few researchers making substantial advances, and the rest of science immediately agreeing. Readers have come to expect this, and therefore one key part of popular science writing is usually an idealized description of the genius and logic behind a new discovery. The other key part is a description of the wonder of nature that has been revealed – wonders are just as crucial to good science writing as they are to tabloid pseudo-science. Greg Myers (1990) thus says that popular science creates a "narrative of nature."

The discrepancy between the idealized version of science common in the media and more messy accounts can sometimes be used for particular purposes. For example, a number of interests are opposed to taking any measures in response to the threat of global warming. In the 1990s, one way in which that opposition was made effective was by questioning the phenomenon itself. Opponents of action in the US Congress and *The Wall Street Journal* – and the phenomenon was much more widespread – drew on the fact that the scientific community did not speak with one voice on the issue. Paul Edwards (1999) argues that every scientific community contains "high-proof" scientists who prioritize empirical evidence and are suspicious of theory and "frontier" scientists who prioritize theory and are willing to consider diverse sources of evidence (we might see this as driven by variation in personality or by some feature of the scientific ecology which creates niches for these attitudes). Opponents of action found high-proof scientists who insisted that the evidence for global warming was incomplete. Given the standard image of science, to many people these critics of global warming appeared more scientific than their colleagues, even though they stood outside of the consensus.

The Dominant Model and its Problems

A number of scholars have recognized a "dominant model" (Hilgartner 1990), "canonical account" (Bucchi 1998), or "diffusionist model" (Lewenstein 1995) of science popularization: Science produces genuine knowledge, but that knowledge is too complicated to be widely understood. Therefore there is a role for mediators who translate genuine scientific knowledge into simplified accounts for general consumption. From the point of view of science, however, simplification always represents distortion. Popularization, then, is a necessary evil, not to be done by working scientists still engaged in productive research – the culture of science heavily discourages scientists from adopting the role of mediators, and shapes how they mediate when they do. Popularization pollutes the sphere of pure research.

As Stephen Hilgartner and Massimiano Bucchi point out, though, popularization of science often feeds back into the research process. In the cold fusion saga, because the claim to have created cold fusion first appeared in the media, most scientists who later became involved in the controversy found out about the claim on the same day as everybody else. The particular way of framing the claim about cold fusion, and something of its cachet, came from media reports (Bucchi 1998; Collins and Pinch 1993). Popularizations affect scientific research because scientists read them, and may even be a large portion of the "popular" audience. Even within specialized fields, scientists may cite articles more often if reports about them have been published in newspapers (Phillips 1991). There is straightforward continuity between real and popular science, though scientists treat them as completely distinct. Popularizations may also affect the shape of scientific research, when they affect public and policy-makers' attitudes toward areas of research: reporting on genetic links to human traits has been extremely widespread, and may have helped to create excitement about genetics, and biotechnology more generally. A more concrete, though unsuccessful, case is Steven Weinberg's *Dreams of a Final Theory* (1992), which appears to be an attempt to argue for building the Superconducting Supercollider, an enormous particle accelerator that was in the process of being canceled because of budgetary pressures.

Bucchi (1998) argues that the dominant model approximately captures what scientists see as appropriate for normal science situations. When disciplinary boundaries are well established and strong, then novel findings and ideas are easily dealt with, either by being incorporated into the discipline or rejected. Popularization follows a standard path, or is easily and clearly labeled as deviant. When disciplinary boundaries are weak, however, scientists may use popular media as an alternative form of communication; they may play out disagreements in the public eye, and even negotiate the science/non-science boundary there. In such cases, disciplinary resources are not enough to resolve conflicts, and so the outside media become more appropriate. Bucchi's thesis is built on very few cases, and may miss examples in which researchers shape fields on the basis of powerful popularizations – Richard Dawkins's *The Selfish Gene* (1976) is an important such example – but it nonetheless appears to capture one of the factors affecting the use of popular media in science.

Jane Gregory and Steve Miller (1998) claim that scientific rules about popularization are often applied self-servingly. Citing Bruce Lewenstein, they claim that these rules "are stressed by scientists who want to criticize or limit other scientists' behavior but are ignored by the same scientists with regard to their own behavior" (Gregory and Miller 1998: 82). In addition "scientists who do not popularize tend to see popularization as something that would damage their own career; however, they also think that other scientists use popularization to advance their careers" (1998: 82–3). Thus the dominant model is a resource used by scientists when it suits their purposes, and ignored when it does not.

The dominant model is a resource for more than just individual scientists, but can be seen as an ideological resource for science as a whole. The notion of popularization as distortion can be used to discredit non-scientists' use of science, reserving the use for scientists. In effect this enlarges the boundaries of the conceptual authority of scientists. This is despite the fact that science depends upon popularization for its authority. If there were no popularizers of any sort, then science would be a much more marginal intellectual activity than it is.

Perhaps most importantly, although scientists routinely complain about simplifications and distortions in popular science, they recognize other forums in which simplification and distortion are acceptable. At some level, most steps in the scientific process involve simplifications: descriptions of techniques are simplified in attempts to universalize them, the complexity of data is routinely simplified in attempts to model it, and so on (Star 1983). Hilgartner (1990) points out that when outside researchers use results from a discipline or problem area, they routinely summarize or reshape those results to fit new contexts. Although particular cases of this reshaping are seen as distortions, in general it is accepted as legitimate. No sharp distinction can be drawn, then, between genuine knowledge and popularization: "Scientific knowledge is constructed through the collective transformation of statements, and popularization can be seen as an extension of this process" (Hilgartner 1990: 522–4). Any statement of scientific knowledge is more or less well suited to its context, and might have to be changed to become suited to some other context, even another scientific context. Popularization is simply a way of moving knowledge into new domains.

Box 16.1   Patient groups

What would later become "Acquired Immune Deficiency Syndrome," or "AIDS," began as "Gay-Related Immune Deficiency." Over the course of the 1980s and 1990s the epidemic was devastating to American gay men, and so AIDS remained – and remains – firmly associated with gay men. Thus the disease hit a group that *was* loosely a community, and one with some experience of activism in defense of rights. The creation of tests for the presence of antibodies to HIV meant that HIV-positive cases could be diagnosed before the onset of any symptoms; many people found themselves, their friends, family, or lovers to be living with a likely death sentence, but able to devote energy to understanding that sentence and the structures around it. As a result, AIDS activists became more effective than analogous activist groups had been. Steven Epstein's (1996) study of the interaction of American AIDS activists and AIDS researchers shows how organized non-scientists can affect not only the questions asked, but the methods used to answer them.

The most obvious face of AIDS activism was the colorful, hard-hitting, and mediagenic political theater most associated with the group ACT UP (the AIDS Coalition to Unleash Power) – the "die-in" is emblematic of ACT UP's protest strategies. More behind the scenes, activists worked within the system by lobbying government agencies and AIDS researchers. To be effective, people working "inside" had to become "lay experts" on AIDS and the research around it. They continually surprised scientists with their sophisticated understandings of the disease, the drugs being explored, the immune system, and the processes of clinical testing. Epstein describes one episode, in which a biostatistician working on AIDS trials, sought out ACT UP/New York's *AIDS Treatment Research Agenda* at a conference: "I walked down to the courtyard and there was this group of guys, and they were wearing muscle shirts, with earrings and funny hair. I was almost afraid ..." But, reading the document, "there were many places where I found it was sensible – where I found myself saying, 'You mean, we're not doing this?' or 'We're not doing it this way?'" (quoted in Epstein 1996: 247). Third, activists took hold of parts of the research and treatment processes themselves. Project Inform in San Francisco did its own clinical trials and epidemiological studies, and groups around the US created buying clubs for drugs.

Among the obvious targets of activism were such things as funding for research and medical treatment. In addition, AIDS activists demanded increased and faster access to experimental drugs. Given the assumption that the disease invariably ended in death, patients demanded the right to decide the level of risk that they were willing to take. To avoid embarrassment, the Food and Drug Administration permitted the administration of these drugs on a "compassionate use" basis.

Perhaps more surprising was a focus on the methodology of clinical trials. After it had been shown that short-term use of azidothymidine (AZT) was effective at slowing the replication of the virus, and was not unacceptably toxic, a second phase of testing was to compare the effects of AZT and a placebo. "In blunt terms, in order to be successful the study required that a sufficient number of patients die: only by pointing to deaths in the placebo group could researchers establish that those receiving the active treatment did comparatively better" (Epstein 1996: 202). Placebos were to remain under attack, and activists were sometimes successful in reshaping research to avoid the use of placebos. Also at issue were simultaneous treatments in clinical trials. Researchers expected subjects to take only the treatment that they were given, whether it was the drug or a placebo. However, patients and their advocates argued that clinical trials should mimic "real-world messiness," and should therefore allow research subjects to go about their lives, including taking alternative treatments (Epstein 1996: 257).

The Public Understanding of Science

We can extend the dominant model of science popularization somewhat, to make it the dominant model of expertise. On that model, scientific and technical literacy is a good in short supply outside the ranks of scientists and engineers. For scientists, this deficit represents a political problem, because the scientifically illiterate are (presumably) less likely to support spending on science, and (presumably) more likely to support measures that constrain research. In addition, given the centrality of science and technology to the modern world, scientific illiteracy is a moral problem, leaving people incapable of understanding the world around them and incapable of acting rationally in that world. Therefore, many people feel that we need more "public understanding of science," and this phrase has come to stand for a movement to teach the public more science.

While there is considerable sympathy within S&TS for the idea that the public should know more about science and technology, S&TS's perspective on expertise leaves it critical of the idea that that should amount simply to teaching people more science (e.g. Locke 2002). As a result, in the context of S&TS the phrase, *public understanding of science* often refers to *studies* of attempts to bring scientific knowledge into the public sphere, and not merely the teaching of science. Here lay reactions to experts are as much of interest as experts' strategies for applying their knowledge.

Because members of the public have pre-existing interests in the problems and their solution, case studies often show a certain level of conflict between lay and scientific understandings. Steven Yearley (1999) summarizes the findings of these case studies in terms of three theses:

1   A large part of the public evaluation of scientific knowledge is via the evaluation of the institutions and scientists presenting that knowledge.
2   Members of the interested public typically have expertise that bears on the problem, which may conflict with scientific expertise.
3   Scientific knowledge contains implicit assumptions about the social world, which members of the public can recognize and with which they can disagree.

That is, scientific knowledge is invariably at least partly tied to the local circumstances of its production. Science is done in highly artificial environments (chapter 15); while these environments support versions of universality and objectivity (chapter 11), they are nonetheless limited. When experts attempt to take science into the public domain, and are confronted by the interested public, those limits can often be seen, especially if the opposition is determined enough (chapter 10).

In these public sphere cases, then, opposition to science is not the result of mere "misunderstandings," but of inadequate scientific work. If a proposed study of or solution to a public sphere problem is not put forward by trustworthy agencies or representatives, if it fails to take account of lay expertise, or makes inadequate sociological assumptions, then it may encounter opposition grounded in legitimate concerns.

Brian Wynne's study of Cumbrian sheep farmers (box 16.2) shows us how these lessons play out in practice (the history of AIDS treatment activism in box 16.1 can be used to make similar lessons). In terms of Yearley's theses:

1   The farmers had a history of muted distrust of government positions and of government scientists on the issue, because these had consistently downplayed dangers, because they appeared to have been covering up problems, and because they made errors about matters on

which they claimed expertise.

2   On a number of occasions scientists ignored the farmers' own expertise about the habits of sheep and the productivity of the hillsides.

3   Scientists made assumptions about the culture and economics of sheep-farming, assumptions that disagreed with the farmers' knowledge of themselves.

The farmers can be seen to have a reasonable response to the scientists, based in a culture that is quite divergent from the scientific one. As we can see again, opposition to science is not the result of a misunderstanding, but is the result of inadequate trust and connections between scientific and lay cultures with very different knowledge traditions.

Box 16.2   Wynne on sheep farming

Following the nuclear accident at Chernobyl in late April 1986, a radioactive cloud passed over much of Northern Europe. Localized rainstorms dumped caesium isotopes over various parts of Britain, particularly at higher elevations, in undetermined amounts. The British government's initial reaction was to dismiss the contamination as negligible, but in mid-June a three-week ban was placed on the movement and slaughter of the affected sheep in one region, Cumbria. At the end of the ban, the restrictions were made indefinite (Wynne 1996).

The scientific advice to the farmers was to do nothing, because the sheep would decontaminate themselves quickly as the caesium sank in the soil and fresh uncontaminated grass grew. It turned out that this advice was based on incorrect assumptions about the soil, only discovered two years later. Meanwhile the farmers' attention turned to the Sellafield Nuclear Plant, which sat approximately in the center of the affected areas. Although there had been uneasiness about Sellafield for years, that uneasiness was countered by community ties to the plant, the largest employer in the area. Now the farmers had reason to suspect that the plant was doing concrete harm. There were many concerns about Sellafield, the main one concerning a 1957 fire, which had emitted a lot of radioactivity. As became clear in the years following Chernobyl, the dangers caused by the Sellafield fire had been covered up, and the fire itself had provided an opportunity to cover up routine emissions of spent fuel. Sellafield had also been at the center of controversies and criticism in the years leading up to Chernobyl, because of elevated rates of leukemia, accusations of illegal discharges, apparently misleading information given to inquiries, and poor safety management.

The sheep farmers developed a thorough skepticism of government positions and those of their scientific representatives. They did not believe the claims of scientists from the Ministry of Agriculture, Fisheries and Food that the caesium "fingerprint" in the Cumbrian hills matched that of Chernobyl rather than Sellafield. Later studies suggested that the farmers were right, that 50 percent of the radioactivity did not come from Chernobyl. When the Ministry advised farmers to keep their lambs in the cleaner valleys longer, that advice was ignored, because the farmers knew that the valleys would become very quickly depleted. An experiment in decontamination involved spreading bentonite at different concentrations in different fields, and fencing in the sheep in those fields; farmers objected that sheep do not thrive when fenced in, but they were ignored, and the experiments were eventually abandoned precisely because the fenced-in sheep were doing poorly.

Success in Public Domains

The social constructivist view brings to the fore the complexity of real-world science and technology – and therefore can potentially contribute to their success (Burningham 1998; Irwin 1995). Successful science in the public sphere can be the result of co-construction of the science and politics; science can more easily solve problems in the public domain if both the problems and the science have been adjusted to fit each other (Shackley and Wynne 1995). Or, successful science can be the result of careful adjustment of scientific knowledge to make it fit the new contexts; assumptions that restrict science to purified domains have to be relaxed, and work has to be done to attune scientific knowledge to the knowledges of others (e.g. Jasanoff and Wynne 1998; Sauer 1998). An exactly analogous argument can be made about technology.

In controversies in the public sphere, the typical model of expertise held by the authorities is one that isolates the scientists as having all, or almost all, of the relevant knowledge. Scientific expertise is seen as universally applicable, because scientific knowledge is universal. As we have seen, there are senses in which scientific knowledge aims at universality. Formal objectivity – the denial of subjectivity – creates a kind of universality. Objective procedures are ones that, if correctly followed by *any* scientist, will produce the same results given the same starting points. The highly abstract character of much scientific knowledge also gives it a kind of universality. Abstractions are valued precisely because they leave aside messy concrete details. To the extent that abstract scientific claims describe the real world, they do not describe *particular* features of it.

Yet this universality of scientific knowledge can also be seen as a kind of locality and particularity. Objectivity and abstraction shape science to particular contexts – the insides of laboratories, highly constructed environments, and kinds of Platonic worlds – albeit contexts that are interesting in their reproducibility or their lack of concrete location. While these interesting shapings allow science to be widely applicable, they also limit its applicability in concrete situations. In addition, scientific knowledge is the result of local and particular processes, the result of network-building in the context of disciplinary and other cultural forces, the result of rhetorical and political work in those same contexts, and sometimes the apparently contingent result of contingencies. Scientific knowledge, then, needs to be seen as quite definitely situated in social and material spaces. Scientific knowledge is, in short, socially constructed.

The dominant model of expertise assumes that science trumps all other knowledge traditions, ignoring claims to knowledge that come out of non-science traditions. For this reason, when scientists find opposition to their claims, they tend to see that opposition as misinformed or even irrational. Frustrating, then, is the fact that controversies between experts and non-experts need not be resolvable in the way that controversies among experts are: mechanisms of closure that are effective within scientific communities may not be effective outside of them. Therefore controversies can be very extended, with lay groups continuing to press issues long after experts have arrived at some consensus (e.g. Martin 1991; Richards 1991).

Scientists can also feel considerable frustration when lawyers challenge their accounts in courts. Especially in the United States, the adversarial system makes no deference to scientific expertise. (Science in the United States is also routinely deconstructed in adversarial political settings.) As a result, scientists on the stand can see their claims systematically deconstructed, and evidence that they see as solid can become suspect and irrelevant – as can be seen vividly in the O. J. Simpson case (see Lynch 1998). The conflict between traditions has led a number of scientists to simply dismiss legal methods as not aimed at the truth and inadequate for evaluating science (e.g. Huber 1994; Koshland 1994). Sheila Jasanoff argues that attention to scientific practice complicates this picture tremendously. While undoubtedly there could be much done by judges and lawyers to bridge the "two cultures" – and Jasanoff makes a number of proposals – it needs to be recognized that the law often reacts to science in ways that are appropriate and useful in its context (Jasanoff 1995; also Dreyfuss 1995). Lawyers understand, perhaps imperfectly, some of the problems with the dominant model of scientific expertise, and their work in courtrooms takes advantage of that understanding.

A View of Science and Technology

S&TS's challenge of our common view of science and technology has some interesting consequences when, as in this chapter, it is brought to bear on the boundaries of scientific knowledge.

The dominant model of popularization assumes that scientific knowledge is not tied to any context. Popularization pollutes pure scientific knowledge by simplifying or otherwise changing it to fit non-scientific contexts. That model ignores the ways in which pure science can be continuous with its popularizations, ways in which "pure" science can even depend upon "popular" science. The model also ignores ways in which science is historically located in disciplinary and other matrixes. Knowledge claims are contextualized and recontextualized within pure science, in ways that scientists understand and accept: a claim appropriate for cancer research may have to be reframed for a physiologist.

The dominant model of scientific expertise, and technical expertise more generally, also assumes that scientific knowledge is not tied to any context. As such, it should apply everywhere and trump all other expertise. That model ignores the ways in which even the most pure of pure science is tied to contexts: namely purified scientific ones. And of course it ignores the many other ways in which science is socially constructed. When science is pulled out of its home contexts, and applied to problems in the public domain, it can fail: fail to win the trust of interested parties, fail to recognize its own social assumptions, and fail to deal adequately with messy features of real-world problems. It takes luck or hard work, of a sort that is both political and technical, to successfully apply scientific and technical expertise to public issues.

These are modest notes on which to end this book, but valuable because they apply some of S&TS's lessons to problems of relationships between general public concerns and science and technology. On that important terrain, S&TS has some important contributions to make.

References

Abraham, Itty 2000: "Postcolonial Science, Big Science, and Landscape." In *Doing Science + Culture*, ed. R. Reid and S. Traweek, pp. 49–70. New York: Routledge.

Ackermann, Robert J. 1985: *Data, Instruments and Theory: A Dialectical Approach to Understanding Science*. Princeton, NJ: Princeton University Press.

Adas, Michael 1989: *Machines as the Measures of Men: Science, Technology, and Ideologies of Western Dominance*. Ithaca, NY: Cornell University Press.

Alder, Ken 1998: "Making Things the Same: Representation, Tolerance and the End of the Ancien Régime in France." *Social Studies of Science* 28: 499–546.

Allison, Paul D. and J. Scott Long 1990: "Departmental Effects on Scientific Productivity." *American Sociological Review* 55: 469–78.

Amann, K. and K. Knorr Cetina 1990: "The Fixation of (Visual) Evidence." In *Representation in Scientific Practice*, ed. M. Lynch and S. Woolgar, pp. 85–122. Cambridge, MA: MIT Press.

Anderson, Warwick 1992: "Where Every Prospect Pleases and Only Man is Vile: Laboratory Medicine as Colonial Discourse." *Critical Inquiry* 18: 506–29.

Ashmore, Malcolm 1989: *The Reflexive Thesis: Wrighting Sociology of Scientific Knowledge*. Chicago: University of Chicago Press.

Ashmore, Malcolm, Michael Mulkay, and Trevor Pinch 1989: *Health and Efficiency: A Sociology of Health Economics*. Milton Keynes: Open University Press.

Ayer, A. J. 1952: *Language, Truth and Logic* (2nd edn., first published 1936). New York: Dover.

Bachelard, Gaston 1984: *The New Scientific Spirit*, tr. A. Goldhammer (first published 1934). Boston: Beacon Press.

Baker, G. P. and P. M. S. Hacker 1984: *Scepticism, Rules, and Language*. Oxford: Blackwell.

Barbercheck, Mary 2001: "Mixed Messages: Men and Women in Advertisements in *Science*." In *Women, Science, and Technology: A Reader in Feminist Science Studies*, ed. M. Wyer, M. Barbercheck, D. Giesman, H. Ö. Öztürk, and M. Wayne, pp. 117–31. New York: Routledge.

Barnes, Barry 1982: *T. S. Kuhn and Social Science*. New York: Columbia University Press.

Barnes, Barry and David Bloor 1982: "Relativism, Rationalism and the Sociology of Knowledge." In *Rationality and Relativism*, ed. M. Hollis and S. Lukes, pp. 21–47. Oxford: Blackwell.

Barnes, Barry, David Bloor, and John Henry 1996: *Scientific Knowledge: A Sociological Analysis*. Chicago: University of Chicago Press.

Barnes, S. B. and R. G. A. Dolby 1970: "The Scientific Ethos: A Deviant Viewpoint." *Archives of European Sociology* 11: 3–25.

Bazerman, Charles 1988: *Shaping Written Knowledge: The Genre and Activity of the Experimental Article in Science*. Madison, WI: University of Wisconsin Press.

Benbow, Camilla P. and Julian C. Stanley 1980: "Sex Differences in Mathematical Ability: Factor or Artifact?" *Science* 220: 1262–4.

Ben-David, Joseph 1991: *Scientific Growth: Essays on the Social Organization and Ethos of Science*, ed. Gad Freudenthal. Berkeley: University of California Press.

Berg, Marc 1997: "Of Forms, Containers, and the Electronic Medical Record: Some Tools for a Sociology of the Formal." *Science, Technology, and Human Values* 22: 403–33.

Berger, Peter L. and Thomas Luckmann 1966: *The Social Construction of Reality: A Treatise in the Sociology of Knowledge*. Garden City, NY: Doubleday.

Bernal, Martin 1987: *Black Athena: The Afro-Asiatic Roots of Classical Civilization*. London: Free Association Books.

Biagioli, Mario 1993: *Galileo, Courtier: The Practice of Science in the Culture of Absolutism*. Chicago: University of Chicago Press.

Biagioli, Mario 1995: "Knowledge, Freedom, and Brotherly Love: Homosociality and the Accademia dei Lincei." *Configurations* 2: 139–66.

Bijker, Wiebe E. 1995: *Of Bicycles, Bakelites, and Bulbs: Toward a Theory of Sociotechnical Change*. Cambridge, MA: MIT Press.

Bimber, Bruce 1994: "Three Faces of Technological Determinism." In *Does Technology Drive History? The Dilemmas of Technological Determinism*, ed. M. R. Smith and L. Marx, pp. 79–100. Cambridge, MA: MIT Press.

Biology and Gender Study Group 1989: "The Importance of Feminist Critique for Contemporary Cell Biology." In *Feminism and Science*, ed. by N. Tuana, pp. 172–87. Bloomington, IN: Indiana University Press.

Bird, Alexander 2000: *Thomas Kuhn*. Princeton, NJ: Princeton University Press.

Bleier, Ruth 1984: *Science and Gender: A Critique of Biology and Its Theories on Women*. New York: Pergamon Press.

Bloor, David 1978: "Polyhedra and the Abominations of Leviticus." *British Journal for the History of Science* 11: 245–72.

Bloor, David 1991: *Knowledge and Social Imagery*, 2nd edn. (first published 1976). Chicago: University of Chicago Press.

Bloor, David 1992: "Left and Right Wittgensteinians." In *Science as Practice and Culture*, ed. A. Pickering, pp. 266–82. Chicago: University of Chicago Press.

Bogen, Jim and Jim Woodward 1988: "Saving the Phenomena." *Philosophical Review* 97: 303–52.

Bogen, Jim and Jim Woodward 1992: "Observations, Theories, and the Evolution of the Human Spirit." *Philosophy of Science* 59: 590–611.

Booth, Wayne C. 1961: *Rhetoric of Fiction*. Chicago: University of Chicago Press.

Bowker, Geoffrey C. 1994: *Science on the Run: Information Management and*

Industrial Geophysics at Schlumberger, 1920–1940. Cambridge, MA: MIT Press.

Bowker, Geoffrey C. and Susan Leigh Star 2000: *Sorting Things Out: Classification and its Consequences*. Cambridge, MA: MIT Press.

Boyd, Richard 1979: "Metaphor and Theory Change: What is 'Metaphor' a Metaphor For?" In *Metaphor and Thought*, ed. A. Ortony, pp. 356–408. Cambridge: Cambridge University Press.

Boyd, Richard N. 1984: "The Current Status of Scientific Realism." In *Scientific Realism*, ed. J. Leplin, pp. 41–82. Berkeley: University of California Press.

Boyd, Richard N. 1985: "Lex Orandi est Lex Credendi." In *Images of Science*, ed. P. Churchland and C. A. Hooker, pp. 3–34. Chicago: University of Chicago Press.

Boyd, Richard 1990: "Realism, Conventionality, and 'Realism About.' " In *Meaning and Method: Essays in Honor of Hilary Putnam*, ed. G. Boolos, pp. 171–95. Cambridge: Cambridge University Press.

Brannigan, Augustine 1981: *The Social Basis of Scientific Discoveries*. Cambridge: Cambridge University Press.

Brante, Thomas and Margareta Hallberg 1991: "Brain or Heart? The Controversy over the Concept of Death." *Social Studies of Science* 21: 389–413.

Breslau, Daniel and Yuval Yonay 1999: "Beyond Metaphor: Mathematical Models in Economics as Empirical Research." *Science in Context* 12: 317–32.

Brewer, William F. and Bruce L. Lambert 2001: "The Theory-Ladenness of Observation and the Theory-Ladenness of the Rest of the Scientific Process." *Philosophy of Science* 68 (Proceedings): S176–S186.

Brighton Women and Science Group 1980: *Alice through the Microscope: The Power of Science over Women's Lives*. London: Virago.

Broad, William and Nicholas Wade 1982: *Betrayers of the Truth: Fraud and Deceit in the Halls of Science*. New York: Simon and Schuster.

Bucchi, Massimiano 1998: *Science and the Media*. London: Routledge.

Bucciarelli, Louis L. 1994: *Designing Engineers*. Cambridge, MA: MIT Press.

Burfoot, Annette 1999: *Encyclopedia of Reproductive Technologies*. Boulder, CO: Westview Press.

Burningham, Kate 1998: "A Noisy Road or Noisy Resident? A Demonstration of the Utility of Social Constructivism for Analysing Environmental Problems." *Sociological Review* 46: 536–63.

Burningham, Kate and Geoff Cooper 1999: "Being Constructive: Social Constructivism and the Environment." *Sociology* 33: 297–316.

Butterfield, Herbert 1931: *The Whig Interpretation of History*. London: Bell.

Callon, Michel 1986: "Some Elements of a Sociology of Translation: Domestication of the Scallops and the Fishermen of St. Brieuc Bay." In *Power, Action and Belief*, ed. J. Law, pp. 196–233. London: Routledge and Kegan Paul.

Callon, Michel 1987: "Society in the Making: The Study of Technology as a Tool for Sociological Analysis." In *The Social Construction of Technological Systems: New Directions in the Sociology and History of Technology*, ed. W. E. Bijker, T. P. Hughes, and T. J. Pinch, pp. 83–103. Cambridge, MA: MIT Press.

Callon, Michel and Bruno Latour 1992: "Don't Throw the Baby Out with the Bath School! A Reply to Collins and Yearley." In *Science as Practice and Culture*, ed. A. Pickering, pp. 343–68. Chicago: University of Chicago Press.

Callon, Michel and John Law 1989: "On the Construction of Sociotechnical Networks: Content and Context Revisited." In *Knowledge and Society, vol. 8: Studies in the Sociology of Science Past and Present*, pp. 57–83. Greenwich, CT: JAI Press.

Callon, Michel and John Law 1995: "Agency and the Hybrid *Collectif.*" *South Atlantic Quarterly* 94: 481–507.

Cambrosio, Alberto, Peter Keating, and Michael Mackenzie 1990: "Scientific Practice in the Courtroom: The Construction of Sociotechnical Identities in a Biotechnology Patent Dispute." *Social Problems* 37: 275–93.

Cambrosio, Alberto, Camille Limoges, and Denyse Pronovost 1990: "Representing Biotechnology: An Ethnography of Quebec Science Policy." *Social Studies of Science* 20: 195–227.

Cantor, G. N. 1975: "A Critique of Shapin's Social Interpretation of the Edinburgh Phrenology Debate." *Annals of Science* 32: 196–219.

Carnap, Rudolf 1952: *The Logical Structure of the World*, tr. R. George (first published 1928). Berkeley: University of California Press.

Carroll-Burke, Patrick 2001: "Tools, Instruments and Engines: Getting a Handle on the Specificity of Engine Science." *Social Studies of Science* 31: 593–626.

Cartwright, Nancy 1983: *How the Laws of Physics Lie*. Oxford: Oxford University Press.

Case, Donald O. and Georgeann M. Higgins 2000: "How Can We Investigate Citation Behavior? A Study of Reasons for Citing Literature in Communication." *Journal of the American Society for Information Science* 51: 635–45.

Casper, Monica J. and Adele E. Clarke 1998: "Making the Pap Smear into the 'Right Tool' for the Job: Cervical Cancer Screening in the USA, circa 1940–95." *Social Studies of Science* 28: 255–90.

Charlesworth, Max, Lyndsay Farrall, Terry Stokes, and David Turnbull 1989: *Life Among the Scientists: An Anthropological Study of an Australian Scientific Community*. Oxford: Oxford University Press.

Churchland, Paul 1988: "Perceptual Plasticity and Theoretical Neutrality: A Reply to Jerry Fodor." *Philosophy of Science* 55: 167–87.

Clark, Tim and Ron Westrum 1987: "Paradigms and Ferrets." *Social Studies of Science* 17: 3–33.

Clarke, Adele 1998: *Disciplining Reproduction: Modernity, American Life Sciences, and "the Problems of Sex."* Berkeley: University of California Press.

Clarke, Adele E. 1990: "A Social Worlds Research Adventure: The Case of Reproductive Science." In *Theories of Science in Society*, ed. S. E. Cozzens and T. F. Gieryn, pp. 15–42. Bloomington, IN: Indiana University Press.

Clause, Bonnie Tocher 1993: "The Wistar Rat as a Right Choice: Establishing Mammalian Standards and the Ideal of a Standardized Mammal." *Journal of the History of Biology* 26: 329–49.

Cockburn, Cynthia 1983: *Brothers: Male Dominance and Technological Change*. London: Pluto Press.

Cockburn, Cynthia 1985: *Machinery of Dominance: Women, Men and Technical Know-how*. London: Pluto Press.

Cole, Jonathan R. 1981: "Women in Science." *American Scientist* 69: 385–91.

Cole, Jonathan R. and Stephen Cole 1973: *Social Stratification in Science*. Chicago and London: University of Chicago Press.

Cole, Jonathan R. and Burton Singer 1991: "A Theory of Limited Differences: Explaining the Productivity Puzzle in Science." In *The Outer Circle: Women in the Scientific Community*, ed. H. Zuckerman, J. R. Cole, and J. T. Bruer, pp. 188–204. New York: W. W. Norton.

Cole, Simon A. 1996: "Which Came First, the Fossil or the Fuel?" *Social Studies of Science* 26: 733–66.

Collins, H. M. 1974: "The TEA Set: Tacit Knowledge and Scientific Networks." *Science Studies* 4: 165–86.

Collins, H. M. 1990: *Artificial Experts: Social Knowledge and Intelligent Machines.* Cambridge, MA: MIT Press.

Collins, H. M. 1991: *Changing Order: Replication and Induction in Scientific Practice,* 2nd edn. (first published 1985). Chicago: University of Chicago Press.

Collins, H. M. 1996: "In Praise of Futile Gestures: How Scientific is the Sociology of Scientific Knowledge?" *Social Studies of Science* 26: 229–44.

Collins, H. M. 1999: "Tantalus and the Aliens: Publications, Audiences, and the Search for Gravitational Waves." *Social Studies of Science* 19: 163–97.

Collins, H. M. and T. J. Pinch 1982: *Frames of Meaning: The Social Construction of Extraordinary Science.* London: Routledge and Kegan Paul.

Collins, H. M. and S. Yearley 1992: "Epistemological Chicken." In *Science as Practice and Culture*, ed. A. Pickering, pp. 301–26. Chicago: University of Chicago Press.

Collins, Harry and Martin Kusch 1998: *The Shape of Actions: What Humans and Machines Can Do.* Cambridge, MA: MIT Press.

Collins, Harry and Trevor Pinch 1993: *The Golem: What Everyone Should Know about Science.* Cambridge: Cambridge University Press.

Collins, Randall 1998: *The Sociology of Philosophies: A Global Theory of Intellectual Change.* Cambridge, MA: Harvard University Press.

Constant, Edward W., II 1984: "Communities and Hierarchies: Structure in the Practice of Science and Technology." In *The Nature of Technological Knowledge*, ed. R. Laudan, pp. 27–46. Dordrecht: D. Reidel.

Cowan, Ruth Schwartz 1983: *More Work for Mother: The Ironies of Household Technology from the Open Hearth to the Microwave.* New York: Basic Books.

Crowe, Michael J. 1988: "Ten Misconceptions about Mathematics and Its History." In *History and Philosophy of Modern Mathematics*, ed. W. Aspray and P. Kitcher, pp. 260–77. Minneapolis: University of Minnesota Press.

Cummiskey, D. 1992: "Reference Failure and Scientific Realism: A Response to the Meta-induction." *British Journal for the Philosophy of Science* 43: 21–40.

Cussins, Charis 1996: "Ontological Choreography: Agency through Objectification in Infertility Clinics." *Social Studies of Science* 26(3): 575–610.

Daston, Lorraine 1991: "Baconian Facts, Academic Civility, and the Prehistory of Objectivity." *Annals of Scholarship* 8: 337–63.

Daston, Lorraine 1995: "The Moral Economy of Science." *Osiris* 10: 3–24.

Daston, Lorraine and Peter Galison 1992: "The Image of Objectivity." *Representations* 40: 83–128.

Davidson, Donald 1974: "On the Very Idea of a Conceptual Scheme." *Proceedings of the American Philosophical Association* 47: 5–20.

Davis, Natalie Zemon 1995: *Women on the Margins: Three Seventeenth-Century Lives*. Cambridge, MA: Harvard University Press.

Dawkins, Richard 1976: *The Selfish Gene*. Oxford: Oxford University Press.

Dear, Peter 1995: *Discipline and Experience: The Mathematical Way in the Scientific Revolution*. Chicago: University of Chicago Press.

Delamont, Sara 1989: *Knowledgeable Women: Structuralism and the Reproduction of Elites*. London: Routledge.

Dewey, John 1929: *The Quest for Certainty: A Study of the Relation of Knowledge and Action*. New York: G. P. Putnam's Sons.

Dobbs, Betty Jo Teeter and Margaret Jacob 1995: *Newton and the Culture of Newtonianism*. New Jersey: Humanities Press.

Dornan, Christopher 1990; "Some Problems in Conceptualizing the Issue of Science and the Media." *Critical Studies in Mass Communication* 7: 48–71.

Dowling, Deborah 1999: "Experimenting on Theories." *Science in Context* 12: 261–74.

Dreyfus, Hubert L. 1972: *What Computers Can't Do: A Critique of Artificial Reason*. New York: Harper and Row.

Dreyfuss, Rochelle Cooper 1995: "Is Science a Special Case? The Admissability of Scientific Evidence After *Daubert v. Merrell Dow*." *Texas Law Review* 73: 1779–1804.

Duncker, Elke 2001: "Symbolic Communication in Multidisciplinary Cooperations." *Science, Technology, and Human Values* 26: 349–86.

Dupré, John 1993: *The Disorder of Things: Metaphysical Foundations of the Disunity of Science*. Cambridge, MA: Harvard University Press.

Easlea, Brian 1986: "The Masculine Image of Science with Special Reference to Physics: How Much does Gender Really Matter?" In *Perspectives on Gender and Science*, ed. J. Harding, pp. 132–59. London: Falmer Press.

Edge, David 1979: "Quantitative Measures of Communication in Science: A Critical Review." *History of Science* 17: 102–34.

Edwards, Paul 1997: *The Closed World: Computers and the Politics of Discourse in Cold War America*. Cambridge, MA: MIT.

Edwards, Paul 1999: "Data-laden Models, Model-filtered Data: Uncertainty and Politics in Global Climate Science." *Science as Culture* 8: 437–72.

Ellul, Jacques 1964: *The Technological Society*, tr. J. Wilkinson. New York: Knopf.

Epstein, Steven 1996: *Impure Science: AIDS, Activism, and the Politics of Knowledge*. Berkeley: University of California Press.

Etzkowitz, Henry, Carol Kemelgor, and Brian Uzzi 2000: *Athena Unbound: The Advancement of Women in Science and Technology*. Cambridge: Cambridge University Press.

Etzkowitz, Henry and Loet Leydesdorff (eds.) 1997: *Universities and the Global Economy: A Triple Helix of University–Industry–Government Relations*. London: Pinter.

Evans, Robert 1997: "Soothsaying or Science? Falsification, Uncertainty and Social Change in Macroeconomic Modelling." *Social Studies of Science* 27: 395–438.

Farley, John and Gerald Geison 1974: "Science, politics and Spontaneous Generation in Nineteenth-century France: The Pasteur–Pouchet Debate." *Bulletin of the History of Medicine* 48: 161–98.

Faulkner, Wendy 2000: "Dualisms, Hierarchies and Gender in Engineering." *Social Studies of Science* 30: 759–92.

Fausto-Sterling, Anne 1985: *Myths of Gender: Biological Theories about Women and Men.* New York: Basic Books.

Fausto-Sterling, Anne 1993: "The Five Sexes: Why Male and Female Are not Enough." *The Sciences* March/April: 20–5.

Findlen, Paula 1993: "Controlling the Experiment: Rhetoric, Court Patronage and the Experimental Method of Francisco Redi." *History of Science* 31: 35–64.

Fine, Arthur 1986: *The Shaky Game: Einstein, Realism and the Quantum Theory.* Chicago: University of Chicago Press.

Fleck, Ludwik 1979: *Genesis and Development of a Scientific Fact* (first published 1935), tr. F. Bradley, T. J. Trenn; ed. T. J. Trenn, and R. K. Merton. Chicago: University of Chicago Press.

Fodor, Jerry 1988: "A Reply to Churchland's 'Perceptual Plasticity and Theoretical Neutrality.' " *Philosophy of Science* 55: 188–98.

Forsythe, Diana E. 2001: *Studying Those Who Study Us: An Anthropologist in the World of Artificial Intelligence,* ed. David J. Hess. Stanford, CA: Stanford University Press.

Fox, Mary Frank. 1983: "Publication Productivity among Scientists: A Critical Review." *Social Studies of Science* 13: 285–305.

Fox, Mary Frank 1991: "Gender, Environmental Milieu, and Productivity in Science." In *The Outer Circle: Women in the Scientific Community,* ed. H. Zuckerman, J. R. Cole, and J. T. Bruer, pp. 188–204. New York: W. W. Norton.

Franklin, Allan 1997: "Calibration." *Perspectives on Science* 5: 31–80.

Franklin, Sarah and Helena Ragoné (eds.) 1998: *Reproducing Reproduction: Kinship, Power, and Technological Innovation.* Philadelphia: University of Pennsylvania Press.

Franklin, Ursula 1990: *The Real World of Technology.* Concord, ON: Anansi.

Fraser, Nancy and Linda J. Nicholson 1990: "Social Criticism without Philosophy: An Encounter between Feminism and Postmodernism." In *Feminism/Postmodernism,* ed. L. J. Nicholson, pp. 19–38. New York: Routledge.

Friedman, Michael 1999: *Reconsidering Logical Positivism.* Cambridge: Cambridge University Press.

Fuchs, Stephan 1992: *The Professional Quest for Truth: A Social Theory of Science and Knowledge.* Albany: State University of New York Press.

Fuchs, Stephan 1993: "Positivism is the Organizational Myth of Science." *Perspectives on Science* 1: 1–23.

Fujimura, Joan H. 1988: "The Molecular Biological Bandwagon in Cancer Research: Where Social Worlds Meet." *Social Problems* 35: 261–83.

Fuller, Steve 2000a: *The Governance of Science: Ideology and the Future of the Open Society.* Buckingham: Open University Press.

Fuller, Steve 2000b: *Thomas Kuhn: A Philosophical History for Our Times.* Chicago: University of Chicago Press.

Gale, George and Cassandra L. Pinnick 1997: "Stalking Theoretical Physicists: An Ethnography Flounders: A Response to Merz and Knorr Cetina." *Social Studies of Science* 27: 113–23.

Galison, Peter 1997: *Image and Logic: A Material Culture of Microphysics*. Chicago: University of Chicago Press.

Galison, Peter and David J. Stump (eds.) 1996: *The Disunity of Science: Boundaries, Contexts, and Power*. Stanford, CA: Stanford University Press.

Garber, Daniel 1995: "Experiment, Community, and the Constitution of Nature in the Seventeenth Century." *Perspectives on Science* 3: 173–205.

Gergen, Kenneth 1986: "Correspondence versus Autonomy in the Language of Understanding Human Action." In *Metatheory in Social Science: Pluralisms and Subjectivities*, ed. D. Fiske and R. Shweder, pp. 132–62. Chicago: University of Chicago Press.

Gibbons, Michael, Camille Limoges, Helga Nowotny, Simon Schwartzmann, Peter Scott and Michael Trow 1994: *The New Production of Knowledge: The Dynamics of Science and Research in Contemporary Societies*. London: Sage.

Gieryn, Thomas F. 1992: "The Ballad of Pons and Fleischmann: Experiment and Narrativity in the (Un)Making of Cold Fusion." In *The Social Dimensions of Science*, ed. E. McMullin, pp. 217–43. Notre Dame, IN: University of Notre Dame Press.

Gieryn, Thomas 1996: "Policing STS: A Boundary-Work Souvenir from the Smithsonian Exhibition on 'Science in American Life.' " *Science, Technology, and Human Values* 21: 100–15.

Gieryn, Thomas F. 1999: *Cultural Boundaries of Science: Credibility on the Line*. Chicago: University of Chicago Press.

Gieryn, Thomas F. and Anne Figert 1986: "Scientists Protect their Cognitive Authority: The Status Degradation Ceremony of Sir Cyril Burt." In *The Knowledge Society: The Growing Impact of Scientific Knowledge on Social Relations*, ed. G. Böhme and N. Stehr, pp. 67–86. Dordrecht: D. Reidel.

Gieryn, Thomas F. and Anne E. Figert 1990: "Ingredients for a Theory of Science in Society: O-Rings, Ice Water, C-Clamp, Richard Feynman, and The Press." In *Theories of Science in Society*, ed. S. E. Cozzens and T. F. Gieryn, pp. 67–97. Bloomington: Indiana University Press.

Gilbert, G. Nigel and Michael Mulkay 1984: *Opening Pandora's Box: A Sociological Analysis of Scientists' Discourse*. Cambridge: Cambridge University Press.

Gilbert, Nigel 1977: "Referencing as Persuasion." *Social Studies of Science* 7: 113–22.

Gilligan, Carol 1982: *In a Different Voice: Psychological Theory and Women's Development*. Cambridge, MA: Harvard University Press.

Gillispie, Charles Coulton 1960: *The Edge of Objectivity: An Essay in the History of Scientific Ideas*. Princeton, NJ: Princeton University Press.

Gingras, Yves and Michel Trépanier 1993: "Constructing a Tokamak: Political, Economic and Technical Factors as Constraints and Resources." *Social Studies of Science* 23: 5–36.

Ginsburg, Faye D. and Rayna Rapp (eds.) 1995: *Conceiving the New World Order: The Global Politics of Reproduction*. Berkeley: University of California Press.

Glover, Judith 2000: *Women and Scientific Employment*. New York: St. Martin's Press.

Goodfield, June 1981: *An Imagined World: A Story of Scientific Discovery*. New York: Harper and Row.

Goodman, Nelson 1983: *Fact, Fiction, and Forecast* (4th edn. first published 1954). Cambridge, MA: Harvard University Press.

Goodwin, Charles 1995: "Seeing in Depth." *Social Studies of Science* 25: 237–74.

Goonatilake, Susantha 1993: "Modern Science and the Periphery." In *The "Racial" Economy of Science*, ed. S. Harding, pp. 259–67. Bloomington: Indiana University Press.

Gould, Stephen Jay 1981: *The Mismeasure of Man*. New York: W. W. Norton.

Gregory, Jane and Steve Miller 1998: *Science in Public: Communication, Culture, and Credibility*. New York: Plenum.

Grint, Keith and Steve Woolgar 1997: *The Machine at Work: Technology, Work and Organization*. Cambridge: Polity Press.

Gross, Alan G. 1990a: "The Origin of Species: Evolutionary Taxonomy as an Example of the Rhetoric of Science." In *The Rhetorical Turn: Invention and Persuasion in the Conduct of Inquiry*, ed. H. W. Simons, pp. 91–115. Chicago: University of Chicago Press.

Gross, Alan G. 1990b: *The Rhetoric of Science*. Cambridge, MA: Harvard University Press.

Grundmann, Reiner and Nico Stehr 2000: "Social Science and the Absence of Nature: Uncertainty and the Reality of Extremes." *Social Science Information* 39: 155–79.

Guice, Jon 1998: "Controversy and the State: Lord ARPA and Intelligent Computing." *Social Studies of Science* 28: 103–38.

Guillory, John 2002: "The Sokal Affair and the History of Criticism." *Critical Inquiry* 28: 470–508.

Guston, David 1999a: "Changing Explanatory Frameworks in the U.S. Government's Attempt to Define Research Misconduct." *Science and Engineering Ethics* 5: 137–54.

Guston, David 1999b: "Stabilizing the Boundary between US Politics and Science: The Rôle of the Office of Technology Transfer as a Boundary Organization." *Social Studies of Science* 29: 87–112.

Hacker, Sally 1990: *Doing it the Hard Way: Investigations of Gender and Technology*. Boston: Unwin Hyman.

Hacking, Ian 1983: *Representing and Intervening: Introductory Topics in the Philosophy of Natural Science*. Cambridge: Cambridge University Press.

Hacking, Ian 1992: "The Self-Vindication of the Laboratory Sciences." In *Science as Practice and Culture*, ed. A. Pickering, pp. 29–64. Chicago: University of Chicago Press.

Hacking, Ian 1999: *The Social Construction of What?* Cambridge, MA: Harvard University Press.

Hanson, Norwood Russell 1958: *Patterns of Discovery: An Inquiry into the Conceptual Foundations of Science*. Cambridge: Cambridge University Press.

Haraway, Donna 1976: *Crystals, Fabrics, and Fields: Metaphors of Organicism in 20th Century Developmental Biology*. New Haven, CT: Yale University Press.

Haraway, Donna 1985: "A Manifesto for Cyborgs: Science, Technology, and Socialist Feminism in the 1980s." *Socialist Review* 80: 65–107.

Haraway, Donna 1988: "Situated Knowledges: The Science Question in Feminism and the Privilege of Partial Perspective." *Feminist Studies* 14: 575–609.

Haraway, Donna 1989: *Primate Visions: Gender, Race, and Nature in the World of Modern Science.* New York: Routledge.

Haraway, Donna J. 1991: *Simians, Cyborgs, and Women: The Reinvention of Nature.* New York: Routledge.

Hård, Mikael 1993: "Beyond Harmony and Consensus: A Social Conflict Approach to Technology." *Science, Technology and Human Values* 18: 408–32.

Harding, Sandra 1986: *The Science Question in Feminism.* Ithaca, NY: Cornell University Press.

Harding, Sandra 1991: *Whose Science? Whose Knowledge? Thinking from Women's Lives.* Ithaca, NY: Cornell University Press.

Harding, Sandra (ed.) 1993: *The "Racial" Economy of Science: Toward a Democratic Future.* Bloomington: Indiana University Press.

Harding, Sandra 1998: *Is Science Multicultural? Postcolonialisms, Feminisms, and Epistemologies.* Bloomington: Indiana University Press.

Harkness, Deborah E. 1997: "Managing an Experimental Household: The Dees of Mortlake and the Practice of Natural Philosophy." *Isis* 88: 247–62.

Hartsock, Nancy C. M. 1983: "The Feminist Standpoint: Developing a Ground for a Specifically Feminist Historical Materialism." In *Feminist Perspectives on Epistemology, Metaphysics, Methodology and Philosophy of Science*, ed. by S. Harding and M. Hintikka. Dordrecht: D. Reidel.

Harwood, Jonathan 1976: "The Race–Intelligence Controversy: A Sociological Approach I – Professional Factors." *Social Studies of Science* 6: 369–94.

Harwood, Jonathan 1977: "The Race–Intelligence Controversy: A Sociological Approach II – 'External' Factors," *Social Studies of Science* 7: 1–30.

Hayles, N. Katherine 1999: *How We Became Posthuman: Virtual Bodies in Cybernetics, Literature, and Informatics.* Chicago: University of Chicago Press.

Headrick, Daniel R. 1988: *The Tentacles of Progress: Technology Transfer in the Age of Imperialism, 1850–1940.* Oxford: Oxford University Press.

Heidegger, Martin 1977: *The Question Concerning Technology, and Other Essays*, tr. W. Lovitt (first published 1954). New York: Harper and Row.

Heilbroner, Robert L. 1994: "Do Machines Make History?" In *Does Technology Drive History? The Dilemma of Technological Determinism*, ed. M. R. Smith and L. Marx; first published 1967, pp. 53–65. Cambridge, MA: MIT Press.

Henke, Christopher R. 2000: "Making a Place for Science: The Field Trial." *Social Studies of Science* 30: 483–511.

Hess, David J. 1997: *Science Studies: An Advanced Introduction.* New York: New York University Press.

Hesse, Mary B. 1966: *Models and Analogies in Science.* Notre Dame, IN: University of Notre Dame Press.

Hilgartner, Stephen 1990: "The Dominant View of Popularization: Conceptual Problems, Political Uses." *Social Studies of Science* 20: 519–39.

Hoffman, Robert R. 1985: "Some Implications of Metaphor for Philosophy and Psychology of Science." In *The Ubiquity of Metaphor: Metaphor in Language and Thought*, ed. W. Paprotté and R. Rirven, pp. 327–80. Amsterdam: John Benjamin Publishing.

Hoffmann, Roald and Shira Leibowitz 1991: "Molecular Mimicry, Rachel and Leah, the Israeli Male, and the Inescapable Metaphor of Science." *Michigan Quarterly*

Review 30: 383–98.

Hubbard, Ruth, Mary Sue Henifin, and Barbara Fried (eds.) 1979: *Women Look at Biology Looking at Women: A Collection of Feminist Critiques.* Cambridge, MA: Schenkman.

Huber, Peter 1994: *Galileo's Revenge: Junk Science in the Courtroom.* New York: Basic Books.

Hughes, Thomas P. 1985: "Edison and Electric Light." In *The Social Shaping of Technology: How the Refrigerator got its Hum,* ed. D. Mackenzie and J. Wajcman pp. 39–52. Milton Keynes: Open University Press.

Hughes, Thomas P. 1987: "The Evolution of Large Technological Systems." In *The Social Construction of Technological Systems,* ed. W. E. Bijker, T. P. Hughes, and T. Pinch, pp. 51–82. Cambridge, MA: MIT Press.

Hull, David L. 1988: *Science as a Process: An Evolutionary Account of the Social and Conceptual Development of Science.* Chicago: University of Chicago Press.

Irwin, Alan 1995: *Citizen Science: A Study of People, Expertise and Sustainable Development.* London: Routledge.

Jacob, Margaret C. 1976: *The Newtonians and the English Revolution, 1689–1720.* Ithaca, NY: Cornell University Press.

Jardine, Nicholas 1991: *Scenes of Inquiry: On the Reality of Questions in the Sciences.* Oxford: Clarendon Press.

Jasanoff, Sheila 1995: *Science at the Bar: Law, Science, and Technology in America.* Cambridge, MA: Harvard University Press.

Jasanoff, Sheila 1996: "Beyond Epistemology: Relativism and Engagement in the Politics of Science." *Social Studies of Science* 26: 393–418.

Jasanoff, Sheila S. 1987: "Contested Boundaries in Policy-relevant Science." *Social Studies of Science* 17: 195–230.

Jasanoff, Sheila and Brian Wynne 1998: "Science and Decisionmaking." In *Human Choice and Climate Change,* vol. 1, ed. S. Rayner and E. L. Malone, pp. 1–87. Columbus, OH: Batelle Press.

Jasper, James M. 1992: "Three Nuclear Energy Controversies." In *Controversy,* 3rd edn. ed. D. Nelkin, pp. 97–111. London: Sage.

Joerges, Bernward 1999: "Do Politics Have Artefacts?" *Social Studies of Science* 29: 411–32.

Jones, Roger S. 1982: *Physics as Metaphor.* Minneapolis: University of Minnesota Press.

Jordan, Kathleen and Michael Lynch 1992: "The Sociology of a Genetic Engineering Technique: Ritual and Rationality in the Performance of the 'Plasmid Prep.'" In *The Right Tools for the Job: At Work in Twentieth Century Life,* ed. A. E. Clark and J. H. Fujimura, pp. 77–114. Princeton, NJ: Princeton University Press.

Kaiser, David 1994: "Bringing the Human Actors Back on Stage: The Personal Context of the Einstein–Bohr Debate." *British Journal for the History of Science* 27: 127–52.

Kaiserfeld, Thomas 1996: "Computerizing the Swedish Welfare State: The Middle Way of Technological Success and Failure." *Technology and Culture* 37: 249–79.

Kay, Lily E. 1995: "Who Wrote the Book of Life? Information and the Transformation of Molecular Biology, 1945–55." *Science in Context* 8: 609–34.

Keller, Evelyn Fox 1983: *A Feeling for the Organism: The Life and Work of Barbara*

McClintock. New York: W. H. Freeman.

Keller, Evelyn Fox 1985: *Reflections on Gender and Science*. New Haven, CT: Yale University Press.

Keller, Evelyn Fox 1992: *Secrets of Life, Secrets of Death: Essays on Language, Gender and Science*. New York: Routledge.

Kennefick, Daniel 2000: "Star Crushing: Theoretical Practice and the Theoreticians' Regress." *Social Studies of Science* 30: 5–40.

Kevles, Daniel J. 1998: *The Baltimore Case: A Trial of Politics, Science, and Character*. New York: W. W. Norton.

Kidder, Tracy 1981: *The Soul of a New Machine*. New York: Avon Books.

Kiernan, Vincent 1997: "Ingelfinger, Embargoes, and Other Controls on the Dissemination of Science News." *Science Communication* 18: 297–319.

Kirkup, Gill and Laurie Smith Keller (eds.) 1992: *Inventing Women: Science, Technology and Gender*. Cambridge: Polity Press.

Kirsch, David A. 2000: *The Electric Vehicle and the Burden of History*. New Brunswick, NJ: Rutgers University Press.

Kitcher, Phillip 1991: "Persuasion." In *Persuading Science: The Art of Scientific Rhetoric*, ed. M. Pera and W. R. Shea, pp. 3–27. Canton, MA: Science History Publications.

Kleinman, Daniel Lee 1998: "Untangling Context: Understanding a University Laboratory in the Commercial World." *Science, Technology, and Human Values* 23: 285–314.

Kline, Ronald 1992: *Steinmetz: Engineer and Socialist*. Baltimore: Johns Hopkins University Press.

Kling, Rob 1992: "Audiences, Narratives, and Human Values in Social Studies of Technology." *Science, Technology and Human Values* 17: 349–65.

Klintman, Mikael 2002: "The Genetically Modified (GM) Food Labelling Controversy: Ideological and Epistemic Crossovers." *Social Studies of Science* 32: 71–91.

Knorr Cetina, K. and A. V. Cicourel (eds.) 1981: *Advances in Social Theory and Methodology: Toward an Integration of Micro- and Macro-sociologies*. Boston: Routledge and Kegan Paul.

Knorr, Karin D. 1977: "Producing and Reproducing Knowledge: Descriptive or Constructive?" *Social Science Information* 16: 669–96.

Knorr, Karin D. 1979: "Tinkering toward Success: Prelude to a Theory of Scientific Practice." *Theory and Society* 8: 347–76.

Knorr Cetina, Karin D. 1981: *The Manufacture of Knowledge: An Essay on the Constructivist and Contextual Nature of Science*. Oxford: Pergamon Press.

Knorr Cetina, Karin D. 1983: "The Ethnographic Study of Scientific Work: Towards a Constructivist Interpretation of Science." In *Science Observed: Perspectives on the Social Study of Science*, ed. by K. D. Knorr Cetina and M. Mulkay, pp. 115–40. London: Sage.

Kohler, Robert E. 1994: *Lords of the Fly: Drosophila Genetics and the Experimental Life*. Chicago: University of Chicago Press.

Koshland, Daniel E. 1994: "Scientific Evidence in Court." *Science* 266: 1787.

Krajewski, Wladyslaw 1977: *Correspondence Principle and Growth of Science*. Dordrecht: D. Reidel.

Kripke, Saul A. 1982: *Wittgenstein on Rules and Private Language*. Cambridge,

MA: Harvard University Press.

Kuhn, Thomas S. 1970a: *The Structure of Scientific Revolutions,* 2nd edn. (first published 1962). Chicago: University of Chicago Press.

Kuhn, Thomas S. 1970b: "Reflections on my Critics." In *Criticism and the Growth of Knowledge,* ed. I. Lakatos and A. Musgrave, pp. 231–78. Cambridge: Cambridge University Press.

Kuhn, Thomas S. 1977: "Second Thoughts on Paradigms." In *The Structure of Scientific Theories,* 2nd edn., ed. F. Suppe, pp. 459–82. Urbana: University of Illinois Press.

Kuklick, Henrika 1991: "Contested Monuments: The Politics of Archaeology in Southern Africa." In *Colonial Situations: Essays on the Contextualization of Ethnographic Knowledge,* ed. G. Stocking Jr., pp. 135–69. Madison, WI: University of Wisconsin Press.

Kula, Witold 1986: *Measures and Men,* tr. R. Szreter. Princeton, NJ: Princeton University Press.

Lakatos, Imre 1976: *Proofs and Refutations: The Logic of Mathematical Discovery,* ed. J. Worrall and E. Zahar. Cambridge: Cambridge University Press.

Laqueur, Thomas 1990: *Making Sex: Body and Gender from the Greeks to Freud.* Cambridge, MA: Harvard University Press.

Latour, Bruno 1983: "Give Me a Laboratory and I Will Raise the World." In *Science Observed: Perspectives on the Social Study of Science,* ed. K. D. Knorr Cetina and M. Mulkay, pp. 141–70. London: Sage Publications.

Latour, Bruno 1987: *Science in Action: How to Follow Scientists and Engineers through Society.* Cambridge, MA: Harvard University Press.

Latour, Bruno 1988: *The Pasteurization of France,* tr. A. Sheridon and J. Law. Cambridge, MA: Harvard University Press.

Latour, Bruno 1990: "The Force and the Reason of Experiment." In *Experimental Inquiries,* ed. H. E. L. Grand, pp. 49–80. Dordrecht: Kluwer.

Latour, Bruno 1993a: *On Technical Mediation: Three Talks prepared for "The Messenger Lectures on the Evolution of Civilization".* Ithaca, NY: Cornell University.

Latour, Bruno 1993b: *We Have Never Been Modern,* tr. C. Porter. New York: Harvester Wheatsheaf.

Latour, Bruno 1999: *Pandora's Hope: Essays on the Reality of Science Studies.* Cambridge, MA: Harvard University Press.

Latour, Bruno and Steve Woolgar 1986: *Laboratory Life: The Construction of Scientific Facts,* 2nd edn. (first published 1979). Princeton, NJ: Princeton University Press.

Laudan, Rachel 1984: "Cognitive Change in Technology and Science." In *The Nature of Technological Knowledge: Are Models of Scientific Change Relevant,* ed. R. Laudan, pp. 83–104. Dordrecht: D. Reidel.

Law, John 1973: "The Development of Specialties in Science: The Case of X-ray Protein Crystallography." *Science Studies* 3: 275–303.

Law, John 1987: "Technology and Heterogeneous Engineering: The Case of Portuguese Expansion." In *The Social Construction of Technological Systems: New Directions in the Sociology and History of Technology,* ed. by W. E. Bijker, T. P. Hughes, and T. J. Pinch, pp. 111–34. Cambridge, MA: MIT Press.

Law, John 1999: "After ANT: Complexity, Naming and Topology." In *Actor Net-*

work Theory and After, ed. J. Law and J. Hassard, pp. 1–14. Oxford: Blackwell.

Layton, Edwin 1971: "Mirror-image Twins: The Communities of Science and Technology in 19th-Century America." *Technology and Culture* 12: 562–80.

Layton, Edwin 1974: "Technology as Knowledge." *Technology and Culture* 15: 31–41.

Le Grand, H. E. 1986: "Steady as a Rock: Methodology and Moving Continents." In *The Politics and Rhetoric of Scientific Method*, ed. J. A. Schuster and R. R. Yeo, pp. 97–138. Dordrecht: D. Reidel.

Leiss, William 1972: *The Domination of Nature*. New York: George Braziller.

Leplin, Jarrett (ed.) 1984: *Scientific Realism*. Berkeley: University of California Press.

Lewenstein, Bruce 1995: "From Fax to Facts: Communication in the Cold Fusion Saga." *Social Studies of Science* 25: 403–36.

Lewontin, Richard 1991: *Biology as Ideology: The Doctrine of DNA*. Toronto: Anansi.

Lightfield, E. Timothy 1971: "Output and Recognition of Sociologists." *The American Sociologist* 6: 128–33.

Locke, David 1992: *Science as Writing*. New Haven and London: Yale University Press.

Locke, David 2002: "The Public Understanding of Science – A Rhetorical Invention." *Science, Technology, and Human Values* 27: 87–111.

Long, J. Scott 2001: "From Scarcity to Visibility: Gender Differences in the Careers of Doctoral Scientists and Engineers." Washington, DC: National Academy Press.

Long, J. Scott and R. McGinnis 1981: "Organizational Context and Scientific Productivity." *American Sociological Review* 46: 422–42.

Longino, Helen E. 1990: *Science as Social Knowledge: Values and Objectivity in Scientific Inquiry*. Princeton, NJ: Princeton University Press.

Lowney, Kathleen S. 1998: "Floral Entrepreneurs: Kudzu as Agricultural Solution and Ecological Problem." *Sociological Spectrum* 18: 93–114.

Lynch, Michael 1985: *Art an Artifact in Laboratory Science: A Study of Shop Work and Shop Talk in a Research Laboratory*. London: Routledge and Kegan Paul.

Lynch, Michael 1990: "The Externalized Retina: Selection and Mathematization in the Visual Documentation of Objects in the Life Sciences." In *Representation in Scientific Practice*, ed. by M. Lynch and S. Woolgar, pp. 153–86. Cambridge, MA: MIT Press.

Lynch, Michael 1991: "Laboratory Space and the Technological Complex: An Investigation of Topical Contextures." *Science in Context* 4: 51–78.

Lynch, Michael 1992a: "Extending Wittgenstein: The Pivotal Move from Epistemology to the Sociology of Science." In *Science as Practice and Culture*, ed. A. Pickering, pp. 215–65. Chicago: University of Chicago Press.

Lynch, Michael 1992b: "From the 'Will to Theory' to the Discursive Collage: A Reply to Bloor's 'Left and Right Wittgensteinians.' " In *Science as Practice and Culture*, ed. A. Pickering, pp. 283–300. Chicago: University of Chicago Press.

Lynch, Michael 1998: "The Discursive Production of Uncertainty: The OJ Simpson 'Dream Team' and the Sociology of Knowledge Machine." *Social Studies of Science* 28: 829–68.

Lynch, William and Ronald Kline 2000: "Engineering Practice and Engineering

Ethics." *Science, Technology, and Human Values* 25: 195–225.

MacArthur, Robert H. and Edward O. Wilson 1967: *The Theory of Island Biogeography*. Princeton, NJ: Princeton University Press.

MacKenzie, Donald 1978: "Statistical Theory and Social Interests: A Case Study." *Social Studies of Science* 8: 35–83.

MacKenzie, Donald 1981: *Statistics in Britain, 1865–1930: The Social Construction of Scientific Knowledge*. Edinburgh: Edinburgh University Press.

MacKenzie, Donald 1989: "From Kwajalein to Armageddon? Testing and the Social Construction of Missile Accuracy." In *The Uses of Experiment: Studies in the Natural Sciences*, ed. D. Gooding, T. Pinch, and S. Schaffer, pp. 409–35. Cambridge: Cambridge University Press.

MacKenzie, Donald 1990: *Inventing Accuracy: A Historical Sociology of Nuclear Missile Guidance*. Cambridge, MA: MIT Press.

MacKenzie, Donald 1999: "Slaying the Kraken: The Sociohistory of a Mathematical Proof." *Social Studies of Science* 29: 7–60.

MacRoberts, M. H. and Barbara R. MacRoberts 1996: "Problems of Citation Analysis." *Scientometrics* 36: 435–44.

McSherry, Corynne 2001: *Who Owns Academic Work? Battling for Control of Intellectual Property*. Cambridge, MA: Harvard University Press.

Maines, Rachel 2001: "Socially Camouflaged Technologies: The Case of the Electromechanical Vibrator." In *Women, Science, and Technology*, ed. M. Wyer, M. Barbercheck, D. Giesman, H. Öztürk, and M. Wayne. New York: Routledge.

Martin, Brian 1991: *Scientific Knowledge in Controversy: The Social Dynamics of the Fluoridation Debate*. Albany, NY: SUNY Press.

Martin, Brian 1996: "Sticking a Needle into Science: The Case of Polio Vaccines and the Origin of AIDS." *Social Studies of Science* 26: 245–76.

Martin, Emily 1991: "The Egg and the Sperm: How Science has Constructed a Romance Based on Stereotypical Male–Female Roles." *Signs* 16: 485–501.

Maynard, Douglas W. and Nora Cate Schaeffer 2000: "Toward a Sociology of Social Scientific Knowledge: Survey Research and Ethnomethodology's Asymmetric Alternates." *Social Studies of Science* 30: 323–70.

Medawar, Peter 1963: "Is the Scientific Paper a Fraud?" *The Listener* 70(1798): 377–8.

Mendelsohn, Everett 1977: "The Social Construction of Scientific Knowledge." In *The Social Production of Scientific Knowledge*, ed. E. Mendelsohn, P. Weingart, and R. Whitley, pp. 3–26. Dordrecht: D. Reidel.

Merchant, Carolyn 1980: *The Death of Nature: Women, Ecology and the Scientific Revolution*. San Francisco: Harper and Row.

Merton, Robert K. 1973: *The Sociology of Science: Theoretical and Empirical Investigations*, ed. N. W. Storer. Chicago: University of Chicago Press.

Merz, Martina and Karin Knorr Cetina 1997: "Deconstruction in a 'Thinking' Science: Theoretical Physicists at Work." *Social Studies of Science* 27: 73–112.

Miettinen, Reijo 1998: "Object Construction and Networks in Research Work: The Case of Research on Cullulose-degrading Enzymes." *Social Studies of Science* 28: 423–64.

Mirowski, Philip 1989: *More Heat than Light: Economics as Social Physics, Physics as Nature's Economics*. Cambridge: Cambridge University Press.

Misa, Thomas J. 1992: "Controversy and Closure in Technological Change: Constructing 'Steel.' " In *Shaping Technology/Building Society: Studies in Sociotechnical Change*, ed. W. Bijker and J. Law, pp. 109–39. Cambridge, MA: MIT Press.

Misak, C. J. 1995: *Verificationism: Its History and Prospects* (1995 edn.). New York: Routledge.

Mitcham, Carl 1994: *Thinking through Technology: The Path between Engineering and Philosophy*. Chicago: University of Chicago Press.

Mitroff, Ian I. 1974: "Norms and Counter-Norms in a Select Group of the Apollo Moon Scientists: A Case Study of the Ambivalence of Scientists." *American Sociological Review* 39: 579–95.

Moore, Kelly 1996: "Organizing Integrity: American Science and the Creation of Public Interest Organizations, 1955–1975." *American Journal of Sociology* 101: 1592–1627.

Moore, Lisa Jean 1997: " 'It's Like You Use Pots and Pans to Cook. It's the Tool': The Technologies of Safer Sex." *Science, Technology, and Human Values* 22: 434–71.

Moravcsik, Michael J. and J. M. Ziman 1975: "Paradisia and Dominatia: Science and the Developing World." *Foreign Affairs* 53: 699–724.

Mukerji, Chandra 1989: *A Fragile Power: Scientists and the State*. Princeton, NJ: Princeton University Press.

Mulkay, Michael 1969: "Some Aspects of Cultural Growth in the Sciences." *Social Research* 36: 22–52.

Mulkay, Michael 1980: "Interpretation and the Use of Rules: The Case of Norms of Science." In *Science and Social Structure: A Festschrift for Robert K. Merton* (Transactions of the New York Academy of Sciences, series II, vol. 39), ed. T. F. Gieryn, pp. 111–25. New York: New York Academy of Sciences.

Mulkay, Michael 1985: *The Word and the World: Explorations in the Form of Sociological Analysis*. London: George Allen and Unwin.

Mulkay, Michael 1989: "Looking Backward." *Science, Technology, and Human Values* 14: 441–59.

Mulkay, Michael 1994: "The Triumph of the Pre-embryo: Interpretations of the Human Embryo in Parliamentary Debate over Embryo Research." *Social Studies of Science* 24: 611–39.

Mumford, Lewis 1934: *Technics and Civilization* (1934 edn.). New York: Harcourt Brace.

Mumford, Lewis 1967: *The Myth of the Machine*, vol. 1. New York: Harcourt Brace Jovanovich.

Myers, Greg 1990: *Writing Biology: Texts in the Social Construction of Scientific Knowledge*. Madison, WI: University of Wisconsin Press.

Myers, Greg 1995: "From Discovery to Invention: The Writing and Rewriting of Two Patents." *Social Studies of Science* 25: 57–106.

Nandy, Ashis 1988: *Science, Hegemony and Violence: A Requiem for Modernity*. Tokyo: United Nations University.

National Academy of Sciences 1995: *On Being a Scientist*. Washington, DC: National Academy of Sciences.

National Science Foundation 1998: *Science and Engineering Indicators 1998*. Arlington, VA: National Science Foundation.

Nelkin, Dorothy 1995: *Selling Science: How the Press Covers Science and Technology*, 2nd edn. New York: W. H. Freeman.

Nelkin, Dorothy and M. Susan Lindee 1995: *The DNA Mystique: The Gene as a Cultural Icon*. New York: W. H. Freeman.

Nersessian, Nancy J. 1988: "Reasoning from Imagery and Analogy in Scientific Concept Formation." In *PSA 1988: Proceedings of the 1988 Biennial Meeting of the Philosophy of Science Association*, vol. 1, ed. A. Fine and J. Leplin, pp. 41–7. East Lansing, MI: Philosophy of Science Association.

Nickles, Thomas 1992: "Good Science as Bad History: From Order of Knowing to Order of Being." In *The Social Dimensions of Science*, ed. E. McMullin, pp. 85–129. Notre Dame, IN: University of Notre Dame Press.

Noble, David F. 1984: *Forces of Production: A Social History of Industrial Automation*. New York: Oxford University Press.

Noble, David F. 1992: *A World Without Women: The Christian Clerical Culture of Western Science*. New York: Alfred A. Knopf.

Nowotny, Helga, Peter Scott, and Michael Gibbons 2001: *Re-Thinking Science: Knowledge and the Public in an Age of Uncertainty*. London: Polity Press.

O'Connell, Joseph 1993: "Metrology: The Creation of Universality by the Circulation of Particulars." *Social Studies of Science* 23: 129–74.

Orenstein, Peggy 1994: *School Girls*. New York: Doubleday.

Oudshoorn, Nelly 1994: *Beyond the Natural Body: An Archaeology of Sex Hormones*. London: Routledge.

Overall, Christine 1993: *Human Reproduction: Principles, Practices, Policies*. Oxford: Oxford University Press.

Packer, Kathryn and Andrew Webster 1996: "Patenting Culture in Science: Reinventing the Scientific Wheel of Credibility." *Science, Technology and Human Values* 21: 427–53.

Papineau, David (ed.) 1996: *The Philosophy of Science*. New York: Oxford University Press.

Pauly, Philip J. 1996: "The Beauty and Menace of the Japanese Cherry Trees: Conflicting Visions of American Ecological Independence." *Isis* 87: 51–73.

Perelman, Chaim and Lucie Olbrechts-Tyteca 1969: *The New Rhetoric: A Treatise on Argumentation*, tr. J. Wilkinson and P. Weaver. Notre Dame, IN: University of Notre Dame Press.

Philip, Kavita 1995: "Imperial Science Rescues a Tree: Global Botanic Networks, Local Knowledge and the Transcontinental Transplantation of Cinchona." *Environment and History* 1: 173–200.

Phillips, David M. 1991: "Importance of the Lay Press in the Transmission of Medical Knowledge to the Scientific Community." *New England Journal of Medicine* (11 October): 1180–3.

Picart, Caroline Joan S. 1994: "Scientific Controversy as Farce: The Benveniste-Maddox Counter Trials." *Social Studies of Science* 24: 7–38.

Pickering, Andrew 1984: *Constructing Quarks: A Sociological History of Particle Physics*. Chicago: University of Chicago Press.

Pickering, Andrew 1992a: "From Science as Knowledge to Science as Practice." In *Science as Practice and Culture*, ed. A. Pickering, pp. 1–26. Chicago: University of Chicago Press.

Pickering, Andrew (ed.) 1992b: *Science as Practice and Culture*. Chicago: University of Chicago Press.

Pickering, Andrew 1995: *The Mangle of Practice: Time, Agency, and Science*. Chicago: University of Chicago Press.

Pinch, Trevor 1985: "Towards an Analysis of Scientific Observation: The Externality and Evidential Significance of Observational Reports in Physics." *Social Studies of Science* 15: 3–36.

Pinch, Trevor 1993a: " 'Testing – One, Two, Three . . . Testing!' Toward a Sociology of Testing." *Science, Technology and Human Values* 18: 25–41.

Pinch, Trevor 1993b: "Turn, Turn, and Turn Again: The Woolgar Formula." *Science, Technology and Human Values* 18: 511–22.

Pinch, Trevor J. and Wiebe E. Bijker 1987: "The Social Construction of Facts and Artifacts: Or How the Sociology of Science and the Sociology of Technology Might Benefit Each Other." In *The Social Construction of Technological Systems: New Directions in the Sociology and History of Technology*, ed. W. E. Bijker, T. P. Hughes, and T. Pinch, pp. 17–50. Cambridge, MA: MIT Press.

Polanyi, Michael 1958: *Personal Knowledge: Towards a Post-critical Philosophy*. Chicago: University of Chicago Press.

Polanyi, Michael 1962: "The Republic of Science: Its Political and Economic Theory." *Minerva* 1: 54–73.

Popper, Karl 1963: *Conjectures and Refutations: The Growth of Scientific Knowledge*. London: Routledge and Kegan Paul.

Porter, Theodore M. 1992a: "Objectivity as Standardization: The Rhetoric of Impersonality in Measurement, Statistics, and Cost–Benefit Analysis." *Annals of Scholarship* 9: 19–59.

Porter, Theodore M. 1992b: "Quantification and the Accounting Ideal in Science." *Social Studies of Science* 22: 633–52.

Porter, Theodore M. 1995: *Trust in Numbers: The Pursuit of Objectivity in Science and Public Life*. Princeton, NJ: Princeton University Press.

Prelli, Lawrence J. 1989: *A Rhetoric of Science: Inventing Scientific Discourse*. Columbia, SC: University of South Carolina Press.

Price, Derek de Solla 1986: *Little Science, Big Science and Beyond* (first published 1963). New York: Columbia University Press.

Proctor, Robert N. 1991: *Value-Free Science? Purity and Power in Modern Knowledge*. Cambridge, MA: Harvard University Press.

Putnam, Hilary 1981: "The 'Corroboration' of Theories." In *Scientific Revolutions*, ed. I. Hacking, pp. 60–79. Oxford: Oxford University Press.

Pyenson, Lewis 1985: *Cultural Imperialism and Exact Sciences: German Expansion Overseas*. New York: Peter Lang.

Radder, Hans 1988: *The Material Realization of Science: A Philosophical View on the Experimental Natural Sciences, Developed in Discussion with Habermas*. Assen/Maastricht: Van Gorcum.

Radder, Hans 1993: "Science, Realization, and Reality: The Fundamental Issues." *Studies in History and Philosophy of Science* 24: 327–49.

Rader, Karen 1998: " 'The Mouse People': Murine Genetics Work at the Bussey Institution, 1909–1936." *Journal of the History of Biology* 31: 327–54.

Rappert, Brian 2001: "The Distribution and Resolutions of the Ambiguities of

Technology, or Why Bobby Can't Spray." *Social Studies of Science* 31: 557–91.

Reardon, Jenny 2001: "The Human Genome Diversity Project: A Case Study in Coproduction." *Social Studies of Science* 31: 357–88.

Restivo, Sal 1990: "The Social Roots of Pure Mathematics." In *Theories of Science in Society*, ed. S. E. Cozzens and T. F. Gieryn, pp. 120–43. Bloomington: Indiana University Press.

Rheinberger, Hans-Jörg 1997: *Toward a History of Epistemic Things: Synthesizing Proteins in the Test Tube*. Stanford, CA: Stanford University Press.

Richards, Evelleen 1991: *Vitamin C and Cancer: Medicine or Politics?* London: Macmillan.

Richards, Evelleen 1996: "(Un)Boxing the Monster." *Social Studies of Science* 26: 323–56.

Richards, Evelleen and John Schuster 1989: "The Feminine Method as Myth and Accounting Resource: A Challenge to Gender Studies and Social Studies of Science." *Social Studies of Science* 19: 697–720

Richardson, Alan W. 1998: *Carnap's Construction of the World: The Aufbau and the Emergence of Logical Empiricism*. Cambridge: Cambridge University Press.

Rose, Hilary 1986: "Beyond Masculinist Realities: A Feminist Epistemology for the Sciences." In *Feminist Approaches to Science*, ed. R. Bleier. New York: Pergamon Press.

Rosen, Paul 1993: "Social Construction of Mountain Bikes: Technology and Postmodernity in the Cycle Industry." *Social Studies of Science* 23: 479–514.

Rossi, Alice 1965: "Women in Science: Why So Few?" *Science* 148, 3674: 1196–1202.

Rossiter, Margaret 1982: *Women Scientists in America: Struggles and Strategies to 1940*. Baltimore: Johns Hopkins University Press.

Rossiter, Margaret 1995: *Women Scientists in America: Before Affirmative Action*. Baltimore: Johns Hopkins University Press.

Roth, Wolff-Michael and G. Michael Bowen 1999: "Digitizing Lizards: The Topology of 'Vision' in Ecological Fieldwork." *Social Studies of Science* 29: 719–64.

Rudwick, Martin J. S. 1974: "Poulett Scrope on the Volcanoes of Auvergne: Lyellian Time and Political Economy." *British Journal for the History of Science* 7: 205–42.

Sauer, Beverly 1998: "Embodied Knowledge: The Textual Representation of Embodied Sensory Information in a Dynamic and Uncertain Material Environment." *Written Communication* 15: 131–69.

Schiebinger, Londa 1993: *Nature's Body: Gender in the Making of Modern Science*. Boston: Beacon Press.

Schiebinger, Londa 1999: *Has Feminism Changed Science?* Cambridge, MA: Harvard University Press.

Scott, Pam, Evelleen Richards, and Brian Martin 1990: "Captives of Controversy: The Myth of the Neutral Social Researcher in Contemporary Scientific Controversies." *Science, Technology, and Human Values* 15: 474–94.

Searle, John R. 1995: *The Construction of Social Reality*. New York: Free Press.

Selzer, Jack (ed.) 1993: *Understanding Scientific Prose*. Madison, WI: University of Wisconsin Press.

Shackley, Simon and Brian Wynne 1995: "Global Climate Change: The Mutual

Construction of an Emergent Science-Policy Domain." *Science and Public Policy* 22: 218–30.

Shapin, Steven 1975: "Phrenological Knowledge and the Social Structure of Early Nineteenth-century Edinburgh." *Annals of Science* 32: 219–43.

Shapin, Steven 1981: "Of Gods and Kings: Natural Philosophy and Politics in the Leibniz–Clarke Disputes." *Isis* 72: 187–215.

Shapin, Steven 1988: "The House of Experiment in Seventeenth-century England." *Isis* 79: 373–404.

Shapin, Steven 1994: *A Social History of Truth: Civility and Science in Seventeenth-century England.* Chicago: University of Chicago Press.

Shapin, Steven and Simon Schaffer 1985: *Leviathan and the Air-Pump: Hobbes, Boyle, and the Experimental Life.* Princeton, NJ: Princeton University Press.

Simon, Bart 1999: "Undead Science: Making Sense of Cold Fusion after the (Arti)fact." *Social Studies of Science* 29: 61–86.

Singleton, Vicky and Mike Michael 1993: "Actor-networks and Ambivalence: General Practitioners in the Cervical Screening Programme." *Social Studies of Science* 23: 227–64.

Sismondo, Sergio 1995: "The Scientific Domains of Feminist Standpoints." *Perspectives on Science* 3: 49–65.

Sismondo, Sergio 1996: *Science without Myth: On Constructions, Reality, and Social Knowledge.* Albany, NY: SUNY Press.

Sismondo, Sergio 1997: "Modelling Strategies: Creating Autonomy for Biology's Theory of Games." *History and Philosophy of the Life Sciences* 19: 147–61.

Sismondo, Sergio 2000: "Island Biogeography and the Multiple Domains of Models." *Biology and Philosophy* 15: 239–58.

Smith, Dorothy 1987: *The Everyday World as Problematic: A Feminist Sociology.* Boston: Northeastern University Press.

Sørensen, Knut H. 1992: "Towards a Feminized Technology? Gendered Values in the Construction of Technology." *Social Studies of Science* 22: 5–31.

Sørensen, Knut H. and Anne-Jorunn Berg 1987: "Genderization of Technology among Norwegian Engineering Students." *Acta Sociologica* 30: 151–71.

Star, Susan Leigh 1983: "Simplification in Scientific Work: An Example from Neuroscience Research." *Social Studies of Science* 13: 205–28.

Star, Susan Leigh 1989: "Layered Space, Formal Representations, and Long-distance Control: The Politics of Information." *Fundamenta Scientiae* 10: 125–54.

Star, Susan Leigh 1991: "Power, Technologies and the Phenomenology of Conventions: On Being Allergic to Onions." In *A Sociology of Monsters: Essays on Power, Technology and Domination*, ed. J. Law, pp. 26–56. London: Routledge.

Star, Susan Leigh 1992: "Craft vs. Commodity, Mess vs. Transcendence: How the Right Tool Became the Wrong One in the Case of Taxidermy and Natural History." In *The Right Tools for the Job: At Work in Twentieth-century Life Sciences*, ed. A. Clarke and J. Fujimura, pp. 257–86. Princeton, NJ: Princeton University Press.

Star, Susan Leigh and James R. Griesemer 1989: "Institutional Ecology, 'Translations' and Boundary Objects: Amateurs and Professionals in Berkeley's Museum of Vertebrate Zoology, 1907–39." *Social Studies of Science* 19: 387–420.

Stepan, Nancy Leys 1986: "Race and Gender: The Role of Analogy in Science."

Isis 77: 261–77.

Stokes, Donald E. 1997: *Pasteur's Quadrant: Basic Science and Technological Innovation*. Washington, DC: Brookings Institution Press.

Takacs, David 1996: *The Idea of Biodiversity: Philosophies of Paradise*. Baltimore: Johns Hopkins University Press.

Taylor, Peter 1995: "Co-Construction and Process: A Response to Sismondo's Classification of Constructivisms." *Social Studies of Science* 25: 348–59.

Tilghman, Shirley 1998: "Science vs. The Female Scientist." In *Scientific Knowledge: Basic Issues in the Philosophy of Science*, ed. J. Kourany, pp. 40–2. Belmont, CA: Wadsworth Publishing.

Toulmin, Stephen 1990: *Cosmopolis: The Hidden Agenda of Modernity*. Chicago: University of Chicago Press.

Traweek, Sharon 1988. *Beamtimes and Lifetimes: The World of High Energy Physicists*. Cambridge, MA: Harvard University Press.

Tribby, Jay 1994: "Club Medici: Natural Experiment and the Imagineering of 'Tuscany'." *Configurations* 2: 215–35.

Tuana, Nancy 1989: "The Weaker Seed: The Sexist Bias of Reproductive Theory." In *Feminism and Science*, ed. by N. Tuana, pp. 147–71. Bloomington, IN: Indiana University Press.

Tufte, Edward 1997: *Visual Explanations: Images and Quantities, Evidence and Narrative*. Cheshire, CT: Graphics Press.

Turkle, Sherry 1984: *The Second Self: Computers and the Human Spirit*. New York: Simon and Schuster.

Turkle, Sherry and Seymour Papert 1990: "Epistemological Pluralism: Styles and Voices within the Computer Culture." *Signs* 16: 128–57.

Turnbull, David 1995: "Rendering Turbulence Orderly." *Social Studies of Science* 25: 9–34.

Turner, S. P. and Chubin, D. E. 1976: "Another Appraisal of Ortega, the Coles and Science Policy: The Ecclesiastes Hypothesis." *Social Science Information* 15: 657–62.

Van den Daele, Wolfgang 1977: "The Social Construction of Science: Institutionalisation and Definition of Positive Science in the Latter Half of the Seventeenth Century." In *The Social Production of Scientific Knowledge*, ed. E. Mendelsohn, P. Weingart, and R. Whitley, pp. 27–54. Dordrecht: D. Reidel.

van Fraassen, Bas C. 1980: *The Scientific Image*. Oxford: Clarendon Press.

Varma, Roli 2000: "Changing Research Cultures in U.S. Industry." *Science, Technology, and Human Values* 25: 395–416.

Vaughan, Diane 1996: *The Challenger Launch Decision: Risky Technology, Culture, and Deviance at NASA*. Chicago: University of Chicago Press.

Vincenti, Walter 1995: "The Technical Shaping of Technology: Real-world Constraints and Technical Logic in Edison's Electrical Lighting System." *Social Studies of Science* 25: 553–74.

Vincenti, Walter G. 1990: *What Engineers Know and How They Know It: Analytical Studies from Aeronautical History*. Baltimore: Johns Hopkins University Press.

Wajcman, Judy 1991: *Feminism Confronts Technology*. University Park, PA: Pennsylvania State University Press.

Weber, Rachel N. 1997: "Manufacturing Gender in Commercial and Military Cockpit Design." *Science, Technology and Human Values* 22: 235–53.

Weinberg, Steven 1992: *Dreams of a Final Theory.* New York: Vintage Books.

Wenneras, Christine and Agnes Wold 2001: "Nepotism and Sexism in Peer-Review." in *The Gender and Science Reader*, ed. M. Lederman and I. Bartsch, pp. 42–8. London: Routledge.

Winner, Langdon 1986: *The Whale and the Reactor: A Search for Limits in an Age of High Technology.* Chicago: University of Chicago Press.

Wittgenstein, Ludwig 1958: *Philosophical Investigations*, 2nd edn., tr. G. E. M. Anscombe. Oxford: Blackwell.

Wood, S. 1982: *The Degradation of Work? Skill, Deskilling and the Labour Process.* London: Hutchinson.

Woolgar, Steve 1981: "Interests and Explanation in the Social Study of Science." *Social Studies of Science* 11: 365–94.

Woolgar, Steve 1988: *Science: The Very Idea.* Chichester, Sussex: Ellis Horwood.

Woolgar, Steve 1992: "Some Remarks about Positionism: A Reply to Collins and Yearley." In *Science as Practice and Culture*, ed. A. Pickering, pp. 327–42. Chicago: University of Chicago Press.

Woolgar, Steve 1993: "What's at Stake in the Sociology of Technology? A Reply to Pinch and to Winner." *Science, Technology and Human Values* 18: 523–9.

Wynne, Brian 1996: "May the Sheep Safely Graze? A Reflexive View of the Expert–Lay Knowledge Divide." In *Risk, Environment and Modernity*, ed. S. Lash, B. Szerszynski, and B. Wynne, pp. 44–83. London: Sage.

Yearley, Steven 1999: "Computer Models and the Public's Understanding of Science: A Case-study Analysis." *Social Studies of Science* 29: 845–66.

Ylikoski, Petri 2001: *Understanding Interests and Causal Explanation.* PhD thesis, University of Helsinki.

Zabusky, Stacia E. and Stephen R. Barley 1997: " 'You Can't be a Stone if You're Cement': Reevaluating the Emic Identities of Scientists in Organizations." *Organizational Behavior* 19: 361–404.

Zenzen, Michael and Sal Restivo 1982: "The Mysterious Morphology of Immiscible Liquids: A Study of Scientific Practice." *Social Science Information* 21: 447–73.

Ziman, John 1984: *An Introduction to Science Studies: The Philosophical and Social Aspects of Science and Technology.* Cambridge: Cambridge University Press.

Ziman, John 1994: *Prometheus Bound: Science in a Dynamic Steady State.* Cambridge: Cambridge University Press.

Zuckerman, Harriet 1977: "Deviant Behavior and Social Control in Science." In *Deviance and Social Change*, ed. E. Sagarin, pp. 87–138. London: Sage Publications.

Zuckerman, Harriet 1984: "Norms and Deviant Behavior in Science." *Science, Technology, and Human Values* 9: 7–13.

Index